Third Generation Solar Cells

This book presents the principle of operation, materials used and possible applications of third generation solar cells that are under investigation and have not been commercialized on a large scale yet. The third generation photovoltaic devices include promising emerging technologies such as organic, dye-sensitized, perovskite and quantum dot-sensitized photocells. This book introduces the reader to the basics of third generation photovoltaics and presents in an accessible way phenomena and a diversity of materials used. In this book, one will find the description of the working principle of new promising solar technologies, their advantages and disadvantages, prospect applications and preliminary analysis of their impact on the environment. The fundamentals of traditional solar cell operation are also included in this book facilitating understanding of new ideas.

This book is ideal reading for everyone who is interested in novel solutions in photovoltaics as well as applications of nanotechnology, photochemistry and materials research.

Third Generation Solar Cells

Agata Zdyb

LONDON AND NEW YORK

First published 2023
by Routledge
Schipholweg 107C, 2316 XC Leiden, The Netherlands
e-mail: enquiries@taylorandfrancis.com
www.routledge.com – www.taylorandfrancis.com

Routledge is an imprint of the Taylor & Francis Group, an informa business

© 2023 Agata Zdyb

The right of Agata Zdyb to be identified as author of this work has been asserted in accordance with sections 77 and 78 of the Copyright, Designs and Patents Act 1988.

All rights reserved. No part of this book may be reprinted or reproduced or utilised in any form or by any electronic, mechanical, or other means, now known or hereafter invented, including photocopying and recording, or in any information storage or retrieval system, without permission in writing from the publishers.

Although all care is taken to ensure integrity and the quality of this publication and the information herein, no responsibility is assumed by the publishers nor the author for any damage to the property or persons as a result of operation or use of this publication and/or the information contained herein.

Library of Congress Cataloging-in-Publication Data
Names: Zdyb, Agata, author.
Title: Third generation solar cells / Agata Zdyb.
Description: Boca Raton : Routledge, Taylor & Francis Group, 2023. | Includes bibliographical references and index.
Identifiers: LCCN 2022030043 (print) | LCCN 2022030044 (ebook) | ISBN 9781032052557 (hbk) | ISBN 9781032052588 (pbk) | ISBN 9781003196785 (ebk)
Subjects: LCSH: Solar cells.
Classification: LCC TK2960 .Z48 2023 (print) | LCC TK2960 (ebook) | DDC 621.31/244—dc23/eng/20220908
LC record available at https://lccn.loc.gov/2022030043
LC ebook record available at https://lccn.loc.gov/2022030044

ISBN: 9781032052557 (hbk)
ISBN: 9781032052588 (pbk)
ISBN: 9781003196785 (ebk)

DOI: 10.1201/9781003196785

Typeset in Times New Roman
by codeMantra

Contents

Preface	*ix*
Author	*xi*

1 Fundamentals of Photovoltaics — 1
- *1.1 Introduction* — *1*
- *1.2 Solar cell: the principle of operation* — *1*
- *1.3 Crystalline silicon solar cells of the first generation* — *6*
  - 1.3.1 The structure of crystalline silicon solar cells — 7
  - 1.3.2 Light absorption — 8
  - 1.3.3 Recombination of carriers — 9
  - 1.3.4 The impact of external conditions on the performance of the solar cell — 9
  - 1.3.5 Recent progress in silicon cells — 11
- *1.4 Thin-film photovoltaic technologies of the second generation* — *12*
  - 1.4.1 CdS/CdTe cell — 14
  - 1.4.2 CIGS solar cell — 16
  - 1.4.3 Silicon thin-film solar cell — 17
  - 1.4.4 GaAs solar cell — 21
- *1.5 Summary* — *22*

2 Organic Solar Cells — 26
- *2.1 Introduction* — *26*
- *2.2 Organic semiconductors* — *26*
- *2.3 Device architecture and performance* — *27*
  - 2.3.1 Single-layer organic photocell — 28
  - 2.3.2 Binary organic photocell — 28
  - 2.3.3 Bulk heterojunction organic photocell — 29
  - 2.3.4 Inverted structure of organic photocell — 34
  - 2.3.5 Ternary blends — 36
- *2.4 Light trapping* — *37*
- *2.5 Stability* — *40*
- *2.6 Summary* — *41*

3 Dye-sensitized Solar Cells — 47
3.1 Introduction — 47
3.2 Structure and basics of DSSC operation — 47
    3.2.1 Work cycle of DSSC — 50
    3.2.2 Inverted configuration of DSSC — 52
3.3 Review of materials used in DSSC — 53
    3.3.1 Mesoporous semiconductor layer — 53
    3.3.2 Sensitizers — 56
    3.3.3 Counter electrode — 59
    3.3.4 Electrolytes — 60
3.4 Summary — 63

4 Perovskite Solar Cells — 69
4.1 Introduction — 69
4.2 Brief history of perovskite photocells — 69
4.3 Operation of perovskite solar cells — 71
    4.3.1 General characteristics of perovskite materials — 71
    4.3.2 Solar cells based on organic-inorganic and all-inorganic perovskites — 76
4.4 Architecture of perovskite cells — 82
    4.4.1 ETL — 83
    4.4.2 HTL — 85
    4.4.3 ETL-free and HTL-free perovskite photocells — 87
4.5 Hysteresis — 88
4.6 Stability — 90
    4.6.1 Thermal stability — 90
    4.6.2 Illumination — 91
    4.6.3 Moisture and oxygen — 92
    4.6.4 The methods of stability improvement — 92
4.7 Lead-free perovskite cells — 94
4.8 Summary — 95

5 Quantum Dot-sensitized Solar Cells — 102
5.1 Introduction — 102
5.2 Structure and operation principle of quantum dot-sensitized solar cells — 102
5.3 Quantum dots as sensitizers — 104
    5.3.1 Doped quantum dots — 105
    5.3.2 Alloy dots — 106
    5.3.3 Core-shell structure — 106
    5.3.4 Co-sensitization — 108
5.4 Multiple exciton generation — 109
5.5 Sensitization process – deposition of quantum dots — 110
    5.5.1 In situ deposition — 110
    5.5.2 Ex situ deposition — 111

	5.6	Other components of QDSSC	113
		5.6.1 Electron transport layer	113
		5.6.2 Electrolyte	114
		5.6.3 Counter electrode	115
	5.7	Summary	115
6	**Environmental Impact of Emerging Photovoltaics**		**119**
	6.1	Introduction	119
	6.2	Life cycle assessment of photovoltaic technologies	119
	6.3	The end-of-life treatment of photovoltaic modules	121
	6.4	Application of life cycle assessment to emerging photovoltaic technologies	123
		6.4.1 Life cycle assessment of organic photovoltaic cells	123
		6.4.2 Life cycle assessment of dye-sensitized photovoltaic cells	127
		6.4.3 Life cycle assessment of perovskite photovoltaic cells	130
		6.4.4 Life cycle assessment of quantum dot-sensitized photovoltaic cells	133
	6.5	Summary	134
7	**Applications of Emerging Photovoltaics – Future Outlook**		**137**
	7.1	Introduction	137
	7.2	Organic photocells	138
	7.3	Dye-sensitized photocells	139
	7.4	Perovskite photocells	140
	7.5	Quantum dot-sensitized photocells	141
	7.6	Indoor applications of emerging photovoltaics	141
	7.7	Tandem photocells	142
	7.8	Summary	144
	Index		149

Preface

Solar radiation is the main resource available on the Earth that can be utilized in all places of the world, independently of geopolitical divisions. Photovoltaic conversion of solar radiation enables direct production of electric energy and contributes to the reduction in harmful emissions resulting from fossil fuels combustion.

Almost 200 years have passed since the discovery of the photovoltaic effect by A.E. Becquerel; however, it is only for 50 years that the intensive development of modern photovoltaics has been observed. Currently, the progress in photovoltaic conversion of solar radiation is driven mainly by the growing energy demand, depletion of fossil fuels and efforts toward sustainability. The observed spread of solar technologies resulted in an increase in the global photovoltaic capacity to around 600 GW. It is expected to rise to 1600 GW in 2030, which will follow the progress in different countries all over the world, mainly China, India, the USA, Germany and Japan. The investigations in the field of photovoltaics resulted in the development of many different kinds of solar cells that are usually classified into three generations: typical wafer-based silicon cells, thin-film technologies based on various materials and emerging photovoltaics that include devices in the research phase, mainly noncommercial.

The proposed book addresses the need for a review of novel photovoltaic technologies belonging to the third generation. This group of photovoltaic devices includes organic photocells, dye-sensitized or quantum dot-sensitized photocells and perovskite photocells. In the proposed book, the fundamentals of photovoltaics are shortly presented first. Then, each main chapter of this book is devoted to the description of operation principle and materials used in the given type of third generation solar device and the progress in this regard. The particular phenomena of physical or chemical nature, fundamental to the emerging photovoltaic cells of each type, are described. This book also covers the topics pertaining to the fabrication of prototype modules, demonstration projects, experimental installations and prospects to enter the market. The stability issue of photocells is discussed, and interesting niche applications, e.g., cells integrated with small indoor devices, are addressed. The life cycle assessment of the solar cells belonging to the third generation is also reviewed in this book.

Author

Agata Zdyb holds the position of Associate Professor at Lublin University of Technology, Poland, and she is the head of the **Department of Renewable Energy Engineering.** She graduated in physics in 1993 from the Faculty of Mathematics, Physics, and Chemistry of the University of Maria Curie-Sklodowska in Lublin and then received her PhD in 2002 from the Gdansk University of Technology in Poland. She completed the habilitation in 2012 at the AGH University of Science and Technology in Cracow, thesis title: "The research on the improvement of dye-sensitized solar cells efficiency".

Agata Zdyb is an author or co-author of over 50 peer-reviewed articles and 5 patent applications. She participated in five scientific projects; currently, she is the member of Reviewer Board for Applied Sciences MDPI journal (IF 2.47) and guest editor for the special issue of Sustainability MDPI journal (IF 2.57). Agata Zdyb was the reviewer of 38 scientific publications. Her scientific topics of interest are thin-film solar cells, dye-sensitized solar cells (DSSC), emerging photovoltaics, nanotechnology for applications in solar cells, photovoltaic systems and renewable energy sources.

Chapter 1

Fundamentals of Photovoltaics

1.1 INTRODUCTION

In a solar cell, the direct conversion of light energy into electric energy takes place via the photovoltaic phenomenon. The essential part of conventional solar cell structure, in which the photovoltaic effect occurs, is the junction formed when an n-type semiconductor and a p-type semiconductor are brought together. The measurement of current-voltage characteristic of the solar cell allows defining the performance parameters of the photocell, such as short-circuit current, open-circuit voltage, fill factor, maximum power point and efficiency of energy conversion. Various semiconductors can be used in solar cells; however, a common choice is silicon, which is a non-toxic, abundant element with bandgap sufficiently matched to solar spectrum. The solar cells based on silicon wafer constitute the first generation of photovoltaic devices, which definitely dominate the photovoltaic market. In the photocells belonging to the second generation, the semiconducting materials such as indium gallium copper diselenide alloys, cadmium telluride or gallium arsenide are used in the form of thin layers.

This chapter presents general topics: the formation of p-n semiconductor junction, the operation of the photovoltaic cell, characteristic parameters and the influence of external conditions on solar cell performance. The description of fundamental phenomena such as light absorption, generation of carriers and recombination is also included. A short overview of photocells classified as both first and second generation is provided. The structure of solar cells of different types and a variety of applied materials are presented, including recent innovations and current record efficiency values.

1.2 SOLAR CELL: THE PRINCIPLE OF OPERATION

A solar cell, also known as photovoltaic cell, is an electrical device that generates electric energy under the influence of illumination. The phenomenon of energy conversion occurs through the photovoltaic effect, a term which is derived from the combination of a Greek word "phos", which means light, and "volt" coming from Alessandro Volta, who invented the electrochemical cell. The photovoltaic (PV) effect takes place in the area of interface created by two materials with different conduction mechanisms. For the first time, it was observed by the French scientist A.E. Becquerel during the illumination of the Pt electrode covered by AgCl or AgBr immersed in an acidic electrolyte.

DOI: 10.1201/9781003196785-1

The first photovoltaic cell, which was intentionally made by the American inventor Charles Fritts, consisted of a thin Se absorber layer deposited on a metal plate and covered by a thin Au film (Green, 2002).

In a typical solar cell, the photovoltaic effect takes place in a p-n semiconductor junction. In the dark, the thermally generated majority carriers which diffuse through the junction in both directions disappear by the recombination with the majority carriers on the other side, which leads to the formation of space charge layers of the remaining dopant ions: positive in the n-type semiconductor and negative in the p-type semiconductor. These regions of space charges are essentially devoid of mobile charge carriers – thus are called depletion regions – and they generate a strong electric field throughout the junction, which prevents further flow of electrons and holes. These processes lead to the occurrence of the electrostatic potential difference between the joined semiconductors, which is called built-in bias. In the n-type semiconductor, the positive electrostatic potential causes a decrease in free electron energy, thus lowering the Fermi level; the analogously generated increase in hole energy in positively charged p-type semiconductor leads to equalization of both Fermi levels (Figure 1.1).

The semiconductors on both sides of the junction region are charge-neutral, and they are referred to as quasi-neutral.

The formatted junction can be described by Poisson's equation (Gray, 2003):

$$\nabla^2 \phi = \frac{q}{\varepsilon}\left(n_0 - p_0 + N_A^- - N_D^+\right), \tag{1.1}$$

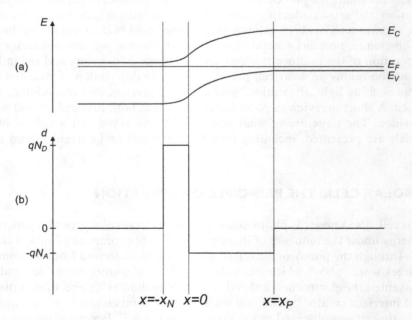

Figure 1.1 The p-n semiconductor junction in equilibrium: (a) energy bands; (b) charge density.

in which ϕ – the electrostatic potential, q – the elementary charge, ε – the electric permittivity of the semiconductor, p_0 – the equilibrium hole concentration, n_0 – the equilibrium electrons concentration, and N_A^-, N_D^+ – the concentration of ionized acceptor and donor, respectively.

Assuming that in the depletion region $p_0 = n_0 = 0$:

$$\nabla^2 \phi = -\frac{q}{\varepsilon} N_D \quad \text{for} \quad -x_N < x < 0, \tag{1.2}$$

$$\nabla^2 \phi = \frac{q}{\varepsilon} N_A, \quad \text{for} \quad 0 < x < x_p. \tag{1.3}$$

In the area of quasi-neutral regions:

$$\nabla^2 \phi = 0, \quad \text{for} \quad x \leq -x_N \quad \text{and} \quad x \geq x_p. \tag{1.4}$$

Upon the absorption of a photon, the energy of which has to be greater than the semiconductor bandgap, the electron is excited to the conduction band and the hole is formed in the valence band. The electrons and holes are then separated by the electric field of ionized dopants in the depletion layer within the junction. The light-generated free electrons in the conduction band are attracted by positively charged dopant ions in the depletion layer, and they diffuse into the n-type semiconductor where they are predominant carriers. Similarly, the positive holes diffuse into the p-type semiconductor. They flow in opposite directions, which leads to the generation of photovoltage between the two sides of the junction.

The total current in the solar cell is expressed as (Gray, 2003):

$$I = I_{SC} - I_{01}\left(\exp\left(\frac{qV}{kT}\right) - 1\right) - I_{02}\left(\exp\left(\frac{qV}{2kT}\right) - 1\right), \tag{1.5}$$

where I_{SC} – the short-circuit current, which consists of the currents in the n type, the p type and the depletion regions, V – the junction voltage, and I_{01}, I_{02} – dark saturation currents due to recombination in quasi-neutral regions (far from the junction) and in the space-charge region, respectively.

Figure 1.2 presents the electric circuit that can be considered as the electric model of an operating solar cell. According to this model, the short-circuit current I_{SC} flows from an ideal source that is connected with two diodes. Diode 1 corresponds to the recombination current in the quasi-neutral regions, and Diode 2 corresponds to the recombination in the depletion region.

At open circuit, the current I_{SC} flows through Diode 1 and the open-circuit voltage is expressed by the following formula (Markvart & Castaňer, 2018):

$$V_{OC} = \frac{kT}{q} \ln \frac{I_{SC} + I_0}{I_0} \approx \frac{kT}{q} \ln \frac{I_{SC}}{I_0}, \tag{1.6}$$

where k – the Boltzmann constant, I_0 – the dark saturation current, and T – the temperature ($I_{SC} \gg I_0$).

The electrical parameters of the solar cell can be determined based on the experimentally obtained current-voltage characteristic that represents the relation between current and voltage under given environmental factors. The measurement of

4 Third Generation Solar Cells

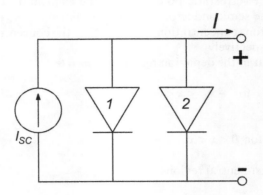

Figure 1.2 Two-diode solar cell circuit model.

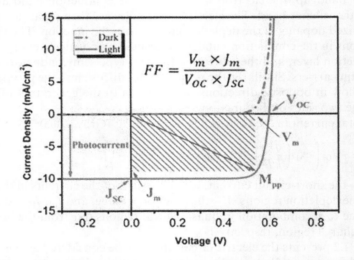

Figure 1.3 Current density (J) vs. voltage (V) characteristic of solar cell in dark and in light. (Copied under Creative Common Attribution 3.0 License from Zaidi (2018).)

I-V curve allows rating maximum power of the photovoltaic cell, short-circuit current and open-circuit voltage, as well as the efficiency. In industries, the fixed set of external conditions called Standard Test Conditions (temperature 25°C, irradiance 1000 W/m² and air mass AM1.5G) is accepted in order to evaluate the performance of photovoltaic modules. The exemplary characteristic of the solar cell is presented in Figure 1.3.

The main characteristic parameters of the solar cell are the following:

- The short-circuit current I_{SC} – the maximum current from the cell occurring at zero voltage, which depends on the intensity and spectrum of incident light, absorption of light and the spatial collection efficiency of carriers.

- The open-circuit voltage V_{OC} – the voltage at zero current flow, decreases with temperature.
- The maximum power P_{MP} – which is a product of the current I_{MP} and voltage V_{MP}.
- Fill factor (FF) – which corresponds to the area of the shadowed rectangle, which is the measure of the quality of the cell and can be calculated according to the following formula:

$$FF = \frac{P_{MP}}{V_{OC}I_{SC}} = \frac{V_{MP}I_{MP}}{V_{OC}I_{SC}}. \tag{1.7}$$

The parameter that is the most commonly used to rate and compare the photovoltaic performance of the cells is the efficiency, which is defined as the fraction of output electrical power to input power of the incident light P_{in}:

$$\eta = \frac{P_{MP}}{P_{in}} = \frac{FF V_{OC} I_{SC}}{P_{in}} \tag{1.8}$$

In practice, the shape of I-V curve differs to some extent from the ideal characteristic, since the real solar cell contains series and shunt (parallel) resistances, which can be included in the two-diode model shown in Figure 1.4.

The series resistance R_s includes all types of bulk and junction resistances in the semiconductors and in metal contacts. It reduces the fill factor and short-circuit current if its value is very high. The shunt resistance, R_{sh}, which results from possible manufacturing defects on the area of photoactive junction, constitutes an additional, internal load to the photocell and thus reduces the photovoltage. In the presence of parasitic resistances, Eq. 1.5 takes the following form (Gray, 2003):

$$I = I'_{SC} - I_{01}\left(\exp\left(q\frac{V+IR_s}{kT}\right)-1\right) - I_{02}\left(\exp\left(q\frac{V+IR_s}{2kT}\right)-1\right) - \frac{V+IR_s}{R_{sh}} \tag{1.9}$$

The effect of both parasitic resistances on the characteristic of the cell is depicted in Figure 1.5.

Another parameter characteristic of the cell, the quantum efficiency (QE), is the ratio of the number of carriers generated to the number of incident photons of certain energy and wavelength. The quantum efficiency would be equal to unity if all photons of a given energy were absorbed and used to generate carriers that reach the electrodes. In fact, the photons of energy lower than the bandgap energy E_g are not

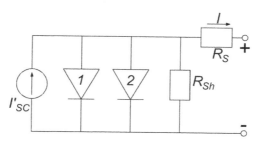

Figure 1.4 Two-diode model of solar cell including parasitic resistances: series resistance R_s and parallel resistance R_{sh}.

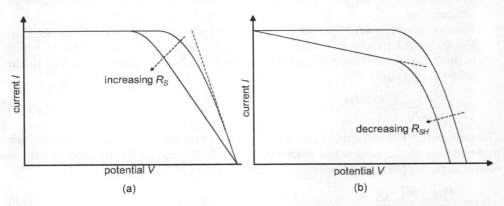

Figure 1.5 The I-V characteristic of a solar cell: (a) effect of series resistance; (b) effect of parallel resistance.

absorbed and the quantum efficiency below the bandgap threshold is reduced due to surface recombination of charge carriers, low diffusion length and optical losses such as reflection and transmission. The quantum efficiency is a function of light wavelength. Taking into account the reflectance and transmittance of the solar cell, it is possible to correct QE and thus obtain the internal quantum efficiency.

The impact of spectral composition of the light on solar cell performance is represented by the spectral response (SR), which is also a function of wavelength λ. The SR is defined as the ratio of generated current to the power of incident radiation. Similar to quantum efficiency, the spectral response can be also limited because of the bandgap width; however, the SR curve drops more quickly toward shorter wavelengths, due to the decrease in the number of photons corresponding to a given light power. The following equation shows the relation between quantum efficiency and spectral response (Dirnberger et al., 2015; Gray, 2003):

$$SR = \frac{q\lambda}{hc} QE \qquad (1.10)$$

where h – the Planck constant, q – the elementary charge, and c – the speed of light.

1.3 CRYSTALLINE SILICON SOLAR CELLS OF THE FIRST GENERATION

The discovery that semiconductor p-n junction generates voltage upon illumination, made in Bell Labs in the USA in 1954, and application of silicon in the first solar cells started the era of photovoltaics. The first produced silicon cell had an efficiency of 6% (Chapin et al., 1954); however, the rapid development in the field of photovoltaics led to a significant increase in the efficiency in the following decades by introducing new materials into the cell structure and modifications of cell configuration. Silicon remains the most important and widely used material in photovoltaics, and silicon cells account for over 80% of global photovoltaic market. In recent years, monocrystalline silicon cells (m-Si) have become popular due to numerous innovations within the modules. Previously, polycrystalline silicon cells (pc-Si) were the most popular ones because of their beneficial ratio of price to efficiency.

1.3.1 The structure of crystalline silicon solar cells

In a typical silicon solar cell, the main part in which the photovoltaic effect takes place consists of the type n and the type p silicon that are placed in contact with each other. In crystalline silicon, every atom makes four covalent bonds with the neighboring atoms. To obtain the n-type silicon, doping with atoms of pentavalent element, e.g., phosphorus, is applied. The atom of a dopant makes four bonds like silicon atom and provides one extra electron. If a trivalent dopant, e.g., boron, is introduced into silicon lattice, it shares only three electrons to make bonds and, in consequence, the vacancy referred to as a hole is generated. In extrinsic silicon type n, which is doped with a donor element, electrons are majority carriers and holes are minority carriers, and in silicon type p, created by doping with an acceptor element, holes are majority carriers and electrons are minority carriers. It is worth emphasizing that a too large concentration of dopants results in bandgap narrowing, which is detrimental to the performance of the cell and therefore has to be avoided. When photons of energy greater than the silicon bandgap value ($E_g = 1.11\,eV$) strike the p-n junction, the photovoltaic effect occurs and the generated charge carriers are moved and separated by the electric field in the junction, which creates voltage between the two sides of the semiconductor. Figure 1.6 shows the scheme of a typical silicon solar cell.

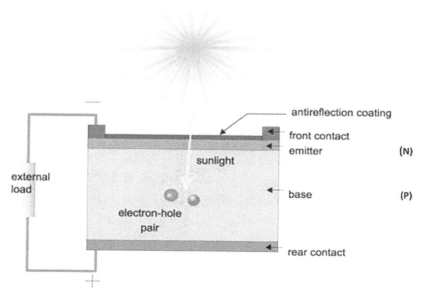

Figure 1.6 The structure of a typical photovoltaic cell. (Copied under Creative Commons Attribution 3.0 License from Zaidi (2018).)

In silicon cells, the front electric contact is a metal grid, usually made of silver. The back contact, which is aluminum, covers the entire surface of the semiconductor.

1.3.2 Light absorption

Photons striking the solar cell can either be absorbed, be reflected or pass through the cell if they have energy lower than the bandgap value. Photon absorption, a phenomenon due to which electron-hole pairs are created, is the only process listed that is beneficial for the performance of a solar cell.

In a semiconductor material, the electron that absorbed the energy of a photon is excited to the conduction band, leaving a hole in the valence band. Two types of semiconductors can be distinguished: direct or indirect bandgap semiconductors, depending on the course of the electron transition. In direct bandgap semiconducting materials such as GaAs, CdTe and Cu(InGa)Se$_2$, the electron transition to the valence band is direct (Figure 1.7a) if parabolic bands are assumed and the absorption coefficient α can be expressed as (Gray, 2003):

$$\alpha(h\nu) \approx A^* \left(h\nu - E_g\right)^{\frac{1}{2}}, \tag{1.11}$$

where ν – the frequency of electromagnetic wave and A^* – a constant.

In indirect bandgap semiconductors such as Si or Ge, phonons appear and take part in the electron transition process, which is shown in Figure 1.7b. The phonon can be either absorbed or emitted, and the absorption coefficient in these two cases is given by:

$$\alpha(h\nu) \approx \frac{B^*}{h\nu} \left(h\nu - E_g\right)^{\frac{3}{2}} \tag{1.12}$$

B^* – a constant.

Reflection of incident light, which reduces the number of potentially absorbed photons, has a negative impact on the performance of the photovoltaic cell. The

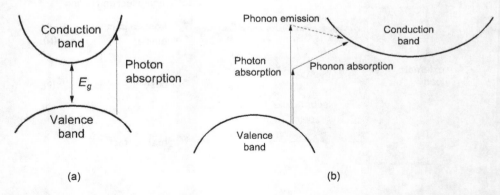

Figure 1.7 Photon absorption in (a) direct bandgap semiconductor and (b) indirect bandgap semiconductor.

coefficient of reflection from the semiconductor surface is described by the following formula (Markvart & Castañer, 2018):

$$R = \frac{(n-1)^2 + k^2}{(n+1)^2 + k^2} \tag{1.13}$$

in which n – the refractive index and k – the extinction coefficient of the semiconductor. In order to limit the reflection of light, two main methods are deployed: antireflective coatings and texturing of front solar cell surface.

Transmission of light through the cell structure is also an undesirable phenomenon. The photons that are lost due to transmission have a long wavelength; their energy is lower than the bandgap value and is partially converted into heat.

1.3.3 Recombination of carriers

The photogenerated pairs of charge carriers last for a certain limited time referred to as the lifetime τ, which is directly related to the diffusion length L, defined as the mean distance that carriers move according to:

$$L = \sqrt{D\tau} \tag{1.14}$$

where D – the diffusion coefficient.

It is desired that both parameters, the lifetime and the diffusion length, are high and influence the increase in the obtained voltage. However, the longer the diffusion length, the more probable the recombination effects in which electron-hole pairs annihilate. There are various types of recombination, depending on the course of the process:

- Radiative recombination – it is a phenomenon opposite to the absorption of light, in which electron returns to the valence band and photon is emitted; it is observed mainly in direct bandgap semiconductors.
- Recombination through traps within the bandgap – the atoms of impurities and defects in crystalline lattice create additional energy levels within the forbidden energy band; the carriers can be trapped and annihilated.
- Auger recombination – the electron returns from conductive band to valence band and emits a quantum, which is then absorbed by another charge carrier that is subsequently excited to the higher energy state.

1.3.4 The impact of external conditions on the performance of the solar cell

The deviations from Standard Test Conditions (STC) in terms of irradiance, temperature and incident light spectrum influence the performance of the solar cell, which means that under real external conditions, generated power and efficiency differ from the nominal values.

The impact of irradiance changes has a complex character. The short-circuit current grows linearly with irradiance, which results in an open-circuit voltage increase due to logarithmic dependence on I_{SC}. However, the final effect on efficiency benefits is limited, since the solar cell is not an ideal source of current, as assumed in the two-diode model. In a real cell, the power is dissipated in parasitic resistances: series and shunt resistances.

As irradiance increases, the series resistance affects the reduction in the efficiency; on the other hand, under decreasing light intensity, the decline in efficiency is an effect of the shunt resistance. The output power of the solar cell is affected by a combination of electric parameters of the cell, and finally, the relation between power and in-plane irradiance G is linear:

$$P_{MP} = G \frac{P_{STC}}{1000 \text{ W/m}^2}, \tag{1.15}$$

where P_{STC} – the power of the cell under STC. The presented relation is not fulfilled for low light intensity in the range of 200–300 W/m² (Louwen et al., 2017).

Photovoltaic cells, as semiconductor devices, are also influenced by temperature changes. According to Eq. 1.6, the open-circuit voltage depends directly on temperature, but dark circuit current present in this formula also varies with temperature, which can be expressed as:

$$I_0 = CT^3 \exp\left(-\frac{E_{g0}}{kT}\right), \tag{1.16}$$

where E_{g0} – the bandgap of the material extrapolated to absolute zero, T – the temperature, and C – a constant, which is independent of temperature. The relation shown in

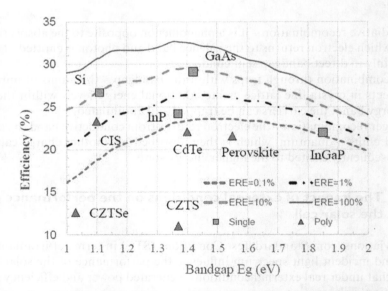

Figure 1.8 Theoretical efficiency limit as a function of bandgap. Record efficiencies for different materials are indicated. (Copied under Creative Commons Attribution 3.0 License from Yamaguchi (2020).)

Eq. 1.16 is critical for the value of V_{OC}, and in consequence, the temperature growth has a negative impact on V_{OC}. However, the role of temperature decreases with the rise in bandgap value. The broader bandgap semiconductors exhibit a better resistance to temperature increase. It is worth noting that the increase in the temperature is also the reason for bandgap narrowing according to the following equation (Markvart & Castaňer, 2018):

$$E_g(T) = E_{g0} - \frac{\alpha T^2}{T + \beta}, \tag{1.17}$$

in which T is the absolute temperature and parameters $\alpha = 4.73 \times 10^{-4}$ eV/K^2, $\beta = 636$ K for Si. In general, the bandgap width in the range of 1.0–1.6 eV is optimal for photovoltaic energy conversion (Figure 1.8) since it enables the best match to the solar spectrum.

1.3.5 Recent progress in silicon cells

Huge efforts have been made over the years in order to improve the performance of the solar cells based on silicon (the bandgap 1.1 eV); however, their efficiency is limited by the maximum theoretical efficiency value determined by Shockley and Queisser. For a single p-n junction Si photovoltaic cell under the illumination of black body spectrum, taking into account only radiative recombination, the upper efficiency is limited to 30% (Shockley & Queisser, 1961; Luque & Marti, 2003). Assuming the more beneficial bandgap width of 1.34 eV and AM1.5G solar spectrum, this value increases to 33.7%.

The scientific research is devoted to overcoming the main problems and unfavorable phenomena observed in practice, which have an adverse influence on the performance of the cells. One of these issues includes optical losses, which occur as a consequence of the reflection of light from the front surface of the cell, as well as spectral mismatch between the solar spectrum and the spectral response. Several routes have been proposed to address the optical losses: light trapping, antireflection coatings and down- and upconversion. Enhancing light absorption by light trapping takes place in textured layers on the front surface of the cell. Pyramid structures or random textures that feature objects of 3–10 µm size in which light bounce back onto the other surface are usually used for this purpose. Another approach to improvement of light absorption, applied especially in thin-film cells, is the introduction of back Bragg mirror – a composition of dielectric layers with different refractive indices that can provide over 90% reflectance (Dubey & Ganesan, 2017). Minimization of reflection from front illuminated surface of the cell is highly desired, and to this aim, also dielectric layers (e.g., TiO_2) with proper thickness and refractive index are used as antireflection coatings. The dielectric, which passivates the surface of the cell, also reduces the number of dangling bonds of Si atoms and limits the surface recombination (Tang et al., 2018).

The spectral mismatch, a reason for the spectral losses, has to be limited to better exploit the solar spectrum. Downconversion is the method of modifying the spectrum of incident light by generation of more than one low-energy photon (visible or NIR) to form a high-energy UV photon. Rare earth ions, which present multiple energy levels, often serve in downconversion. For example, Tb^{3+} absorbs at 488 nm (2.53 eV) and transfers energy to two Yb^{3+} ions, which emit two photons at 980 nm (1.26 eV).

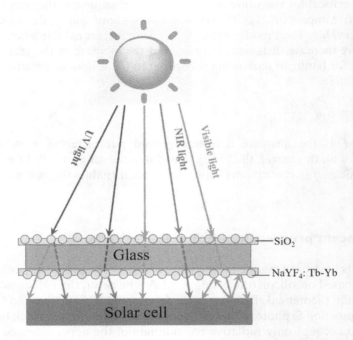

Figure 1.9 The downconvertors placed in front of the solar cell. (Reprinted from Yao and Tang (2020). Copyright with permission from Elsevier.)

The downconvertors are placed in front of the cell, and they can be introduced in the antireflection coating as depicted in Figure 1.9. An example of such material is Tb^{3+}-Yb^{3+} co-doped $NaYF_4$ phosphors, which can be added in the SiO_2 sol-gel to prepare anti-reflection coatings (Yao & Tang, 2020). Other rare earth ions, such as Pr^{3+}, Yb^{3+} or Bi^{3+}, Nd^{3+}, can also be used in luminescent phosphors (Zhou et al., 2019; Li et al., 2018).

Upconversion processes, also beneficial for the performance of the photocells, which shift long wavelength photons to near-infrared or visible range, have also been reported in trivalent lanthanides, especially in Er^{3+}, Ho^{3+} and Yb^{3+}, Tm^{3+} (Venkata Krishnaiah et al., 2017; Rodríguez-Rodríguez et al., 2016). Upconvertors should be placed at the back side of the cell, e.g., at the rear of bifacial silicon cells (Lahoz et al., 2011).

1.4 THIN-FILM PHOTOVOLTAIC TECHNOLOGIES OF THE SECOND GENERATION

Thin-film photovoltaic technology from the beginning aroused the interest due to the low consumption of materials active in conversion of light, low weight of modules, fewer processing steps than Si technology and the large-scale capability. In the years 2007–2011, the market share of thin-film technologies in total global PV production exceeded 10% reaching 17% in 2009 (Photovoltaics Report, 2021), but then the role of thin-film modules decreased and the market was dominated by economically viable multi-Si technology. Since 2016, a gradual decline in the share of multi-Si and poly-Si cells in favor of monocrystalline silicon has been observed, due to numerous innovations improving the price-efficiency ratio of mono-Si technology. Recently, a clear growth in the efficiency of thin-film cells has resulted in a renewed interest and increase in annual global production of thin-film modules. The thin-film technologies gaining popularity include cadmium telluride (CdTe), which reached the laboratory cell efficiency of up to 21%, and copper gallium indium diselenide (CIGS) with highest achieved efficiency of 23.35% (Green et al., 2021). The performance parameters of thin-film solar cells have significantly improved over the years. The current values of photovoltaic parameters provided by thin-film cells and minimodules measured under STC and confirmed by certified laboratories are presented in Table 1.1.

A standard thin-film module consists of glass covered by back electrode layer, semiconductor window/absorber junction and transparent front electrode. The upper layer of the semiconductor junction (window) has a large bandgap and transmits the light to a lower part (absorber) where the main part of charge generation and separation occurs. The described sequence of the layers constitutes the so-called substrate configuration; however, the cell structure can also be deposited in reverse order, starting with front contact (Aliyu et al., 2012). The latter case results in the superstrate configuration, in which the light enters the cell through the glass or polymer substrate and passes transparent conductive oxide (TCO), which makes front contact layer as shown in Figure 1.10. The substrate configuration can be obtained on different kinds of non-transparent foils, e.g., steel, molybdenum or polyamide; therefore, it is applicable in flexible cells. For the superstrate configuration, the subsequent layers are usually deposited on low-cost soda lime glass that is sufficiently transparent (Kirchartz et al., 2016).

The thickness of the layers constituting the thin-film module ranges from a few nanometers to tens of micrometers. Diverse semiconductor materials can be applied in thin-film cells, such as cadmium telluride, copper gallium indium diselenide, silicon in amorphous, micromorph or polycrystalline form, as well as different $A^{III}B^{V}$ compounds, which are used mainly in tandem, multi-junction configurations. Each type of

Table 1.1 The photovoltaic parameters provided by thin-film cells and minimodules measured under STC and confirmed by certified laboratories

Technology	J_{SC} (mA/cm^2)	V_{OC} (V)	FF (%)	PCE (%)
Si amorphous cell	16.36	0.896	69.8	10.2
Si thin-film minimodule	29.7	0.492	72.1	10.5
CIGS cell	39.58	0.734	80.4	23.35
CIGSSe submodule	37.63	0.688	75.8	19.6
CdTe cell	30.25	0.8759	79.4	21.0
GaAs cell	29.78	1.1272	86.7	29.1

Source: Data from Green et al. (2021).

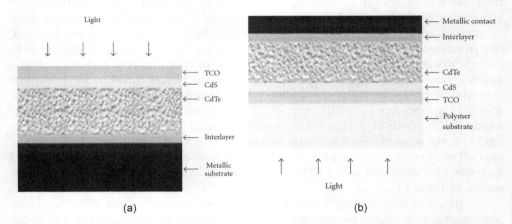

Figure 1.10 Structure of CdTe cell: (a) substrate configuration; (b) superstrate configuration. (Copied under Creative Commons Attribution License from Aliyu et al. (2012).)

semiconductor presents its own characteristic spectral response. At long wavelengths, the spectral response is limited by the bandgap value of the given material, since the photons of energy lower than the bandgap are not absorbed.

In thin-film solar cells, the thickness of the absorber layer is lower than the optical absorption length; therefore, light trapping methods have to be applied in the cells of this type. The light trapping issue is addressed by introducing textured surfaces, grating nanostructures, plasmonic nanostructures or back reflector layers (Amalathas & Alkaisi, 2019).

1.4.1 CdS/CdTe cell

The CdS/CdTe modules, the production of which reached 5.7 GWp in 2019, are the most successful thin-film photovoltaic devices on the market (Photovoltaics Report, 2021). The advantages of this technology include an excellent match of CdTe absorber bandgap to the solar spectrum, as well as fast and reliable methods of thin-layer deposition. The CdTe has a direct bandgap of 1.45 eV and an absorption coefficient in the order of 10^5cm^{-1} (better than Si), which allows 90% absorption at a few micrometers thickness (Romeo, 2018); however, it is worth emphasizing that the properties of thin films depend on the type of the substrate and the parameters of the growth process, e.g., temperature.

In the case of efficient CdTe solar cells, the sub-micrometer thickness of absorber layer is required to provide suitable structural and electrical properties. The structure of CdTe is zinc blende. The common methods of fabricating the absorber layer are vapor transport deposition, close space sublimation, electrodeposition, pulsed laser deposition, metal organic chemical vapor deposition, hot wall epitaxy, colloidal solution and thermal evaporation spray pyrolysis and magnetron sputtering

(He et al., 2021; Razi et al., 2020; Ojo & Dharmadasa, 2016; Arce-Plaza et al., 2017). The CdS with a large bandgap of 2.42 eV is a window layer, which transmits most of the incident light to the absorber. The fabrication of CdS layer is realized by a simple chemical bath technique or vacuum evaporation, and the film thickness used in practice can vary in the range of 50–100 nm (Ashok et al., 2020). The CdS grows naturally as n-type material and CdTe as p-type; therefore, the formation of p-n junction is natural; however, the obtained structure needs chemical treatment with $CdCl_2$ and annealing. This additional procedure improves the quality of the absorber and enhances the efficiency of the cells. The popular methods such as X-ray diffraction, scanning electron microscopy and atomic force microscopy confirmed recrystallization, removing the defects and grain size increase in polycrystalline CdTe upon $CdCl_2$ treatment. Other methods such as X-ray photoelectron spectroscopy and Raman spectroscopy proved the intermixing and reduction in lattice mismatch at the CdS/CdTe interface (Dharmadasa, 2014). Additionally, the residues of Cl in CdTe reduce the electric resistance.

The fabrication of a stable back electric contact with CdTe is a challenge due to very high electron affinity of this material. In fact, the Schottky barrier is present at the junction of CdTe with metal, which can be lowered by additional doping of CdTe surface and introduction of a buffer layer with high carrier concentration. A low-resistance contact can be provided by copper; however, metals such as Cu or Zn are known for diffusion in semiconductor structure that leads to shunting the CdS/CdTe junction. In order to produce reliable contacts, other materials such as Sb, or Au enriched by nanostructures were applied and tested (Masood et al., 2020; Zhu et al., 2020).

The material serving as a front, illuminated electric contact should assure very good transparency and high conductivity. Different transparent conductive oxides such as indium tin oxide (ITO), fluorine-doped tin oxide (FTO) or aluminum-doped ZnO layers of 200–400 nm thickness and conductivity around 10 ohm/sq are usually used in this role (Huang et al., 2019; He et al., 2021; Bittau et al., 2018; Hossain et al., 2018).

In order to lower cost of the CdS/CdTe cells production and to avoid problems with Te scarcity, by reducing the amount of materials, an ultrathin CdTe absorber can be applied. This approach enables the fabrication of semitransparent cells which can utilize the light incoming from front and rear of the cell and find applications in building integrated photovoltaics. In order to achieve the optimum performance of semitransparent CdTe cells, both semiconductor thin films and the layers for electric contact have to be properly fabricated. The best proven method of absorber deposition in this case is magnetron sputtering enabling to control the working gas pressure, power of magnetron and substrate temperature (Islam et al., 2017). The obtained ultrathin CdTe layer is very sensitive to standard $CdCl_2$ post-deposition treatment, so the time of this procedure has to be optimized. The ultrathin p-n junction has to be accompanied by transparent contact layers to achieve an effect of bifacial power generation by the cell. Therefore, ZnTe:Cu, copper and carbon nanostructures or CuCl on ITO glass are usually employed as a back contact material (Huang et al., 2019). The current-volatge characteristic and quantum efficiency of the exemplary ultrathin CdTe cell, with a 12 nm thick CuCl back contact is presented in Figure 1.11.

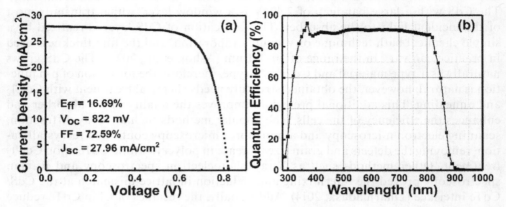

Figure 1.11 Current-voltage characteristic (a) and quantum efficiency (b) of the CdTe thin-film solar cell. (Reprinted from Huang et al. (2019). Copyright (2019) with permission from Elsevier.)

1.4.2 CIGS solar cell

The family of copper chalcopyrites Cu(In, Ga, Al)(Se, S)$_2$ exhibit the broad range of bandgap energies 1.04–2.4 eV matching to spectrum of visible light. Among these compounds, CuInSe$_2$ and CuGaSe$_2$ are the components of indium gallium copper diselenide Cu(In, Ga)Se$_2$ quaternary semiconductor alloy, which crystallizes in the structure of tetragonal chalcopyrite. The bandgap of the initially used CuInSe$_2$ in solar cells was extended from 1.04 eV to the optimal value of around 1.15 eV by the inclusion of Ga atoms, which also improved the electronic properties of the absorber material (Rau & Schock, 2018). The precise adjustment of the bandgap is possible by making changes in the composition of CuIn$_{1-x}$Ga$_x$Se$_2$ (CIGS) alloy, which is determined by mole fraction of Ga content $x = $ Ga/(Ga+In). The relation between the x and bandgap value is expressed by the following equation (Shafarman & Stolt, 2003):

$$E_g = 1.010 + 0.626x - 0.167x(1-x). \tag{1.18}$$

The change of x from 0 to 1 implies variation of E_g from 1.04 to 1.64 eV (Bouabdelli et al., 2020). Similar to CdTe, CIGS has a direct bandgap and a high absorption coefficient of 10^5 cm^{-1} (Kato, 2017).

The CIGS absorber can be obtained by the co-evaporation process of the elements with the excess of Se, in which the changes of deposition rate and temperature enable controlling the steps of the process and the final composition of the layer. Another method of absorber fabrication is sputtering of metals (Cu, In and Ga) followed by selenization, which takes place in a H$_2$Se atmosphere or rapid thermal process in Se atmosphere. Regardless of the method, the goal is to obtain the high-quality p-type material with the carrier concentration of 10^{16}/cm^3 (Vidal Lorbada et al., 2019). The growth of CdS n-type film and formation of junction are usually realized by chemical bath deposition. The metal of choice for the back contact is molybdenum; however, the available alternatives which can be deposited by electron beam evaporation are W, Cr, Nb, Ti and Mn (Bouabdelli et al., 2020; Orgassa et al., 2003). A key factor in the production of efficient CIGS cells is the application of soda lime glass as the substrate for the back contact. The Na atoms diffuse to CIGS layer, and their incorporation improves the morphology and conductivity as the result of various overlapping effects (Rau & Schock, 2018). The interface between CIGS and metal back contact is a place where deleterious recombination occurs. Therefore, it is beneficial to introduce highly doped layer of SnS, Al_2O_3 or PbS on the back side of absorber to create back surface field (BSF), which reduces the recombination (Barman & Kalita, 2021). Due to BSF, the enhancement of efficiency can be achieved without increase of absorber thickness.

The front contact in the CIGS cell is TCO grown by chemical vapor deposition, magnetron sputtering or atomic layer deposition (Arepalli et al., 2021; Shen et al., 2018). For economic reasons and problems with indium scarcity, the material for front contact is usually ZnO doped with Al, grown in magnetron sputtering system using the $ZnO:Al_2O_3$ ceramic target (Lee et al., 2018; Zdyb et al., 2018). Due to the transparency exceeding 80% in Vis-NIR range, resistivity below 10^{-3} $\Omega \cdot cm$ and bandgap over 3.5 eV, ZnO:Al is a valuable substitute for ITO. The typical structure of CIGS thin-film photocell is shown in Figure 1.12.

1.4.3 Silicon thin-film solar cell

Thin-film photovoltaic cells based on silicon combine the long-term experience with mature, highly efficient silicon technology and low material usage in automated

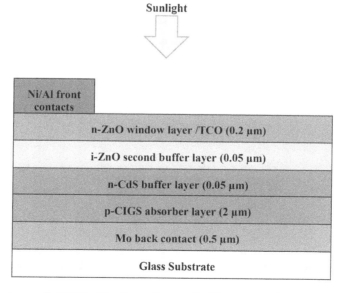

Figure 1.12 Structure of CIGS thin-film photocell. (Reprinted from Bouabdelli et al. (2020). Copyright with permission from Elsevier.)

production of thin-film modules. For many years, amorphous silicon (a-Si) has played an important role among thin-film technologies with a share of 9% in the total global PV production in 2000. In recent years, the a-Si share dropped from around 4% in 2011 to 0.2% in 2019 due to the low efficiency of a-Si modules and the significant enhancement of the performance of other thin-film technologies, including the development of efficient thin crystalline and polycrystalline silicon solar cells (Photovoltaics Report, 2021).

Thin layers of hydrogenated amorphous silicon (a-Si:H) are used in a-Si cells. Hydrogen atoms passivate most of the Si dangling bonds occurring in disordered amorphous material, but still a high density of defects, which act as recombination centers, reduces the electronic quality of the material. Another problematic issue is the Staebler-Wronski effect, which is initial photodegradation of a-Si cells upon exposure to light, due to increased number of mid-gap defects (Shah, 2018). The partial limitation of photodegradation and the overall improvement of the performance was achieved owing to the introduction of hydrogen atoms and the decrease in a-Si film thickness as well as the application of intrinsic layer between p- and n-doped Si (the p-i-n structure is shown in Figure 1.13). The intrinsic layer, characterized by a higher carrier lifetime than the doped material, is the place where the absorption of light occurs and the photogenerated carriers are separated by an internal electric field. The cell can be fabricated in superstrate configuration as well as in the substrate configuration, where the n-i-p structure is deposited on the substrate covered by the back contact layer.

The drawback of a-Si is also too high optical gap of around 1.75 eV, which limits the absorption range. The development and application of hydrogenated microcrystalline silicon with a bandgap of 1.12 eV enabled replacing a-Si and achieving the efficiency of the single microcrystalline Si cells (μc-Si:H) reaching up to 11.9% (Kirchartz

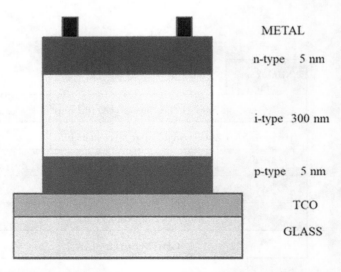

Figure 1.13 Scheme of thin-film p-i-n solar cell. (Copied under Creative Commons Attribution 3.0 License from Gradišnik and Gumbarević (2018).)

et al., 2016; Green et al., 2021). However, the disadvantages of microcrystalline Si, such as low absorption coefficient, high amorphous phase fraction and photodegradation linked to adsorption of oxygen, prevent the µc-Si:H module production. The µc-Si:H cell is used as a bottom subcell in the tandem structure with a-Si:H called micromorph tandem. The recent modification is to substitute µc-Si:H by nanocrystalline Si, which provides the efficiency of 12.7% in the tandem cell (Green et al., 2021).

The development of methods for the fabrication of thin crystalline silicon layers applied together with a reduction in the reflection of incident light led to the obtaining of the modern types of silicon cells with a thickness lower than 50 µm. Thin crystalline silicon films can be produced by different procedures (Mauk, 2018):

- Thinning of the silicon wafer, which is bonded to the substrate (glass or steel) or occurs in the free-standing form.
- Direct deposition of crystalline Si thin films on low-cost substrates by standard methods, e.g., chemical vapor deposition, plasma spraying and liquid phase epitaxy.

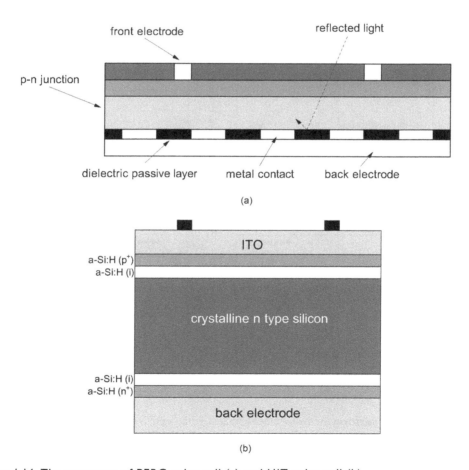

Figure 1.14 The structure of PERC solar cell (a) and HIT solar cell (b).

- The growth of thin layers of silicon on the porous silicon substrate, separation and transferring of Si film to another material, e.g., flexible substrate.

The latter method is applied in the production of heterojunction with intrinsic thin layer (HIT), which consists of a thin monocrystalline Si wafer and layers of doped and intrinsic amorphous silicon on front and bottom sides, as shown in Figure 1.14. The same approach is applied in the production of passivated emitter and rear contact (PERC) Si modules. Both HIT and PERC (Figure 1.14) commercial modules have the efficiency over 20% (Dullweber et al., 2020).

Thin silicon wafers with a thickness of 200–500 μm can also be used in hybrid cells with transparent front PEDOT:PSS (poly(3,4-ethylenedioxythiophene)-polystyrene sulfonate) layer transporting holes (Gao et al., 2021). The introduction of PEDOT:PSS led to an efficient improvement of light absorption by the cell and a power conversion efficiency of up to 12.35%. Figure 1.15 shows the current density vs. voltage characteristics under AM 1.5G solar spectrum and, in dark, EQE and IQE curves.

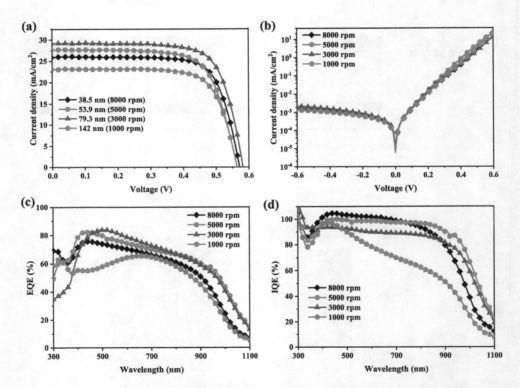

Figure 1.15 The performance of PEDOT:PSS/Si solar cell with different PEDOT:PSS film thicknesses of 38.5–142 nm obtained for different spin coating speeds of 1000–8000 rpm. Current density vs. voltage characteristics (a) under AM 1.5G solar spectrum (b) in dark. EQE curve (c). IQE curve (d). (Reprinted from Gao et al. (2021). Copyright with permission from Elsevier.)

1.4.4 GaAs solar cell

With an optimal direct bandgap (1.44 eV) for light harvesting and resistance to high temperature and radiation, gallium arsenide is a perfect material for solar cells applied in space technologies. A variety of heterostructures (AlGaAs/GaAs, GaInP/GaAs and GaAs/Ge), especially multi-junction combinations (Figure 1.16), exhibit excellent performance. Both mechanically stacked and monolithic multi-junction cells reach the efficiency of well over 30% without concentrators of light, under STC terrestrial conditions (Green et al., 2021). However, the fabrication of photovoltaic devices containing $A^{III}B^{V}$ compounds is expensive and energy intense, even if low-cost Si substrate is employed, because it requires multilayer heterostructures with tunnel junction or mis-oriented substrates or buffer layers prepared in a special way (Dawidowski et al., 2021; Seredin et al., 2018).

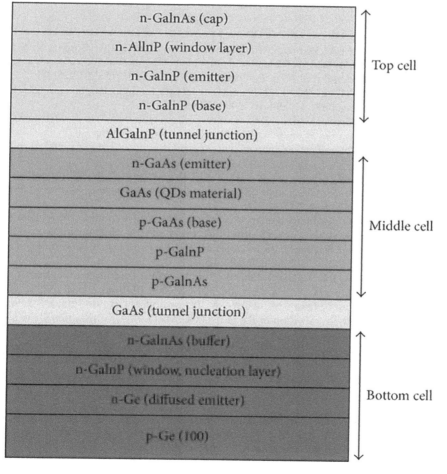

Figure 1.16 Structure of multi-junction solar cell. (Copied under Creative Commons Attribution License from Turala et al. (2013).)

1.5 SUMMARY

According to the adopted general classification of the photovoltaic cells, monocrystalline and polycrystalline silicon wafer-based cells belong to the first generation, while the second generation includes the photocells based on thin films of other semiconductors, e.g., indium gallium copper diselenide alloys, cadmium telluride, amorphous silicon and gallium arsenide. The energy conversion taking place in photocells, based on the photovoltaic phenomenon, includes the following stages: absorption of light, generation of electron-hole pairs, separation and transport of charge carriers. The course of the photovoltaic effect is influenced by a variety of factors, including the type of semiconductor used and external conditions. In order to improve the efficiency of light-to-electricity conversion, different modifications in the solar cell structure and the materials used are introduced. The current record efficiencies of solar cells reach over 25% for silicon cells and 29% for single-junction thin-film GaAs cells.

REFERENCES

Aliyu, M.M., Islam, M.A., Hamzah, N.R., Karim, M.R., Matin, M.A., Sopian, K. & Amin, N. (2012) Recent developments of Flexible CdTe solar cells on metallic substrates: issues and prospects. *International Journal of Photoenergy* 351381. doi: 10.1155/2012/351381.

Amalathas, A.P. & Alkaisi, M.M. (2019) nanostructures for light trapping in thin film solar cells. *Micromachines* 10(9), 619.doi: 10.3390/mi10090619.

Arce-Plaza, A., Andrade-Arvizu, J.A., Courel, M., Alvarado, J.A. & Ortega-López, M. (2017) Study and application of colloidal systems for obtaining CdTe+Te thin films by spray pyrolysis. *Journal of Analytical and Applied Pyrolysis* 124, 285–289. doi: 10.1016/j.jaap.2017.01.022.

Arepalli, K.V., Lee, W., Chung, Y. & Kim, J. (2021) Growth and device properties of ALD deposited ZnO films for CIGS solar cells. *Materials Science in Semiconductor Processing* 121, 105406. doi: 10.1016/j.mssp.2020.105406.

Ashok, A., Regmi, G., Romero-Núñez, A., Solis-López, M., Velumani, S. & Castaneda, H. (2020) Comparative studies of CdS thin films by chemical bath deposition techniques as a buffer layer for solar cell applications. *Journal of Materials Science: Materials in Electronics* 31, 7499–7518. doi: 10.1007/s10854-020-03024-3.

Barman, B. & Kalita, P.K. (2021) Influence of back surface field layer on enhancing the efficiency of CIGS solar cell. *Solar Energy* 216, 329–337. doi: 10.1016/j.solener.2021.01.032.

Bittau, F., Potamialis, C., Togay, M., Abbas, A., Isherwood, P.J.M., Bowers, J.W.& Walls, J.M. (2018) Analysis and optimisation of the glass/TCO/MZO stack for thin film CdTe solar cells. *Solar Energy Materials and Solar Cells* 187, 15–22. doi: 10.1016/j.solmat.2018.07.019.

Bouabdelli, M.W., Rogti, F., Maache, M. & Rabehi, A. (2020) Performance enhancement of CIGS thin-film solar cell. *Optik* 216, 164948. doi: 10.1016/j.ijleo.2020.164948.

Chapin, D., Fuller, C. & Pearson, G. (1954) A new silicon p-n junction photocell for converting solar radiation into electrical power. *Journal of Applied Physics* 25(5), 676–677. doi: 10.1063/1.1721711.

Dawidowski, W., Ściana, B., Zborowska-Lindert, J., Mikolášek, M., Kováč, J., Tłaczała, M. (2021) Tunnel junction limited performance of InGaAsN/GaAs tandem solar cell. *Solar Energy* 214, 632–641. doi: 10.1016/j.solener.2020.11.067.

Dharmadasa, I.M. (2014) Review of the $CdCl_2$ treatment used in CdS/CdTe thin film solar cell development and new evidence towards improved understanding. *Coatings* 4(2), 282–307. doi: 10.3390/coatings4020282.

Dirnberger, D., Blackburn, G., Müller, B. & Reise, C. (2015) On the impact of solar spectral irradiance on the yield of different PV technologies. *Solar Energy Materials and Solar Cells* 132, 431–442. doi: 10.1016/j.solmat.2014.09.034.

Dubey, R.S. & Ganesan, V. (2017) Fabrication and characterization of TiO_2/SiO_2 based Bragg reflectors for light trapping applications. *Results in Physics* 7, 2271–2276. doi: 10.1016/j.rinp.2017.06.041.

Dullweber, T., Stöhr, M., Kruse, C., Haase, F., Rudolph, M., Beier, B., Jäger, P., Mertens, V., Peibst, R. & Brendel, R. (2020) Evolutionary PERC+ solar cell efficiency projection towards 24% evaluating shadow-mask-deposited poly-Si fingers below the Ag front contact as next improvement step. *Solar Energy Materials and Solar Cells* 212, 110586. doi: 10.1016/j.solmat.2020.110586.

Gao, Z., Gao, T., Geng, Q., Lin, G., Li, Y., Chen, L. & Li, M. (2021) Improving light absorption of active layer by adjusting PEDOT:PSS film for high efficiency Si-based hybrid solar cells. *Solar Energy* 228, 299–307. doi: 10.1016/j.solener.2021.09.064.

Gradišnik, V. & Gumbarević, D. (2018) a-Si:H p-i-n photodiode as a biosensor. In: Chee, K. (ed.) *Advances in Photodetectors: Research and Applications.* IntechOpen: London. doi: 10.5772/intechopen.80503.

Gray, J.L. (2003) The physics of the solar cell. In: Luque, A. & Hegedus, S. (eds.) *Handbook of Photovoltaic Science and Engineering.* Wiley: Hoboken, NJ, pp. 61–112.

Green, M.A. (2002) Photovoltaic principles. *Physica E* 14(1–2), 11–17. doi: 10.1016/S1386-9477(02)00354-5.

Green, M., Dunlop, E., Hohl-Ebinger, J., Yoshita, M., Kopidakis, N. & Hao, X. (2021) Solar cell efficiency tables (version 59). *Progress in Photovoltaics* 30(1), 3–12. doi: 10.1002/pip.3506.

He, F., Li, J., Lin, S., Long, W., Wu, L., Hao, X., Zhang, J. & Feng, L. (2021) Semitransparent CdTe solar cells with $CdCl_2$ treated absorber towards the enhanced photovoltaic conversion efficiency. *Solar Energy* 214, 196–204. doi: 10.1016/j.solener.2020.11.049.

Hossain, M.S., Rahman, K.S., Islam, M.A., Akhtaruzzaman, M., Misran, H., Alghoul, M.A. & Amin, N. (2018) Growth optimization of ZnxCd1-xS films on ITO and FTO coated glass for alternative buffer application in CdTe thin film solar cells. *Optical Materials* 86, 270–277. doi: 10.1016/j.optmat.2018.09.045.

Huang, J., Yang, D., Li, W., Zhang, J., Wu, L. & Wang, W. (2019) Copassivation of polycrystalline CdTe absorber by CuCl thin films for CdTe solar cells. *Applied Surface Science* 484, 1214–1222. doi: 10.1016/j.apsusc.2019.03.253.

Islam, M.A., Rahman, K.S., Sobayel, K., Enam, T., Ali, A.M., Zaman, M., Akhtaruzzaman, M. & Amin, N. (2017) Fabrication of high efficiency sputtered CdS:O/CdTe thin film solar cells from window/absorber layer growth optimization in magnetron sputtering. *Solar Energy Materials and Solar Cells* 172, 384–393. doi: 10.1016/j.solmat.2017.08.020.

Kato, T. (2017) $Cu(In, Ga)(Se, S)_2$ solar cell research in Solar Frontier: progress and current status 2017. *Japanese Journal of Applied Physics* 56(4S), 04CA02. https://iopscience.iop.org/article/10.7567/JJAP.56.04CA02.

Kirchartz, T., Abou-Ras, D. & Rau, U. (2016) Introduction to thin-film photovoltaics. In: Abou-Ras, D., Kirchartz, T. & Rau, U. (eds.) *Advanced Characterization Techniques for Thin Film Solar Cells*, 2nd edition. Wiley-VCH Verlag GmbH & Co: Weinheim, Germany, pp. 3–40.

Lahoz, F., Pérez-Rodríguez, C., Hernández, S.E., Martín, I.R., Lavín, V. & Rodríguez-Mendoza, U.R. (2011) Upconversion mechanisms in rare-earth doped glasses to improve the efficiency of silicon solar cells. *Solar Energy Materials and Solar Cells* 95(7), 1671–1677. doi: 10.1016/j.solmat.2011.01.027.

Lee, W.-J., Cho, D.-H., Wi, J.-H., Han, W.S., Kim, B.-K., Choi, S.D., Baek, J.-Y. & Chung, Y.-D. (2018) Characterization of bilayer AZO film grown by low-damage sputtering for Cu(In, Ga)Se_2 solar cell with a CBD-ZnS buffer layer. *Materials Science in Semiconductor Processing* 81, 48–53. doi: 10.1016/j.mssp.2018.03.008.

Li, J., Zhang, S., Luo, H., Mu, Z., Li, Z., Du, Q., Feng, J. & Wu, F. (2018) Efficient near ultraviolet to near infrared downconversion photoluminescence of La_2GeO_5: Bi^{3+}, Nd^{3+} phosphor for silicon-based solar cells. *Optical Materials* 85, 523–530. doi: 10.1016/j.optmat.2018.09.024.

Louwen, A., de Waal, A.C., Schropp, R.E.I., Faaij, A.P.C. & van Sark, W.G.J.H.M. (2017) Comprehensive characterization and analysis of PV module performance under real operating conditions. *Progress of Photovoltaics Research and Applications* 25, 218–232. doi: 10.1002/pip.2848.

Luque, A. & Marti, A. (2003) Theoretical limits of photovoltaic conversion. In: Luque, A. & Hegedus, S. (eds.) *Handbook of Photovoltaic Science and Engineering*. Wiley: Hoboken, NJ, pp. 113–149.

Markvart, T. & Castañer, L. (2018) Semiconductor materials and modelling. In: Kalogirou, S.A. (ed.) *McEvoy's Handbook of Photovoltaics: Fundamentals and Applications*, 3rd edition. Academic Press: Cambridge, MA, pp. 29–54.

Masood, H.T., Anwer, S., Wang, X., Ali, A. & Deliang, W. (2020) Carbon-nanofibers film as a back-contact buffer layer in CdTe thin film solar cell. *Optik* 224, 165505. doi: 10.1016/j.ijleo.2020.165505.

Mauk, M.G. (2018) Thin crystalline and polycrystalline silicon solar cells. In: Kalogirou, S.A. (ed.) *McEvoy's Handbook of Photovoltaics: Fundamentals and Applications*, 3rd edition. Academic Press, pp. 159–226.

Ojo, A.A. & Dharmadasa, I.M. (2016) 15.3% efficient graded bandgap solar cells fabricated using electroplated CdS and CdTe thin films. *Solar Energy* 136, 10–14. doi: 10.1016/j.solener.2016.06.067.

Orgassa, K., Schock, H.W. & Werner, J.H. (2003) Alternative back contact materials for thin film Cu(In, Ga)Se$_2$ solar cells. *Thin Solid Films* 431–432, 387–391. doi: 10.1016/S0040-6090(03)00257-8.

Photovoltaics Report (2021) Fraunhofer Institute for Solar Energy Systems. Available from: https://www.ise.fraunhofer.de/content/dam/ise/de/documents/publications/studies/Photovoltaics-Report.pdf [Accessed December 15, 2021].

Rau, U. & Schock, H.-W. (2018) Cu(In, Ga)Se$_2$ thin-film solar cells. In: Kalogirou, S.A. (ed.) *McEvoy's Handbook of Photovoltaics: Fundamentals and Applications*, 3rd edition. Academic Press: Cambridge, MA, pp. 378–418.

Razi, S.A., Das, N.K., Farhad, S.F.U. & Matin, M.A. (2020) Influence of the $CdCl_2$ solution concentration on the properties of CdTe thin films. *International Journal of Renewable Energy Research* 10(2), 1013–1018. https://www.ijrer.org/ijrer/index.php/ijrer/article/view/10887/pdf.

Rodríguez-Rodríguez, H., Imanieh, M.H., Lahoz, F. & Martín, I.R. (2016) Analysis of the upconversion process in Tm^{3+} doped glasses for enhancement of the photocurrent in silicon solar cells. *Solar Energy Materials and Solar Cells* 144, 29–32. doi: 10.1016/j.solmat.2015.08.017.

Romeo, A. (2018) CdTe solar cells. In: Kalogirou, S.A. (ed.) *McEvoy's Handbook of Photovoltaics: Fundamentals and Applications*, 3rd edition. Academic Press: Cambridge, MA, pp. 309–358.

Seredin, P.V., Lenshin, A.S., Zolotukhin, D.S., Arsentyev, I.N., Zhabotinskiy, A.V. & Nikolaev, D.N. (2018) Impact of the substrate misorientation and its preliminary etching on the structural and optical properties of integrated GaAs/Si MOCVD heterostructures. *Physica E: Low-Dimensional Systems and Nanostructures* 97, 218–225. doi: 10.1016/j.physe.2017.11.018.

Shafarman, W.N. & Stolt, L. (2003) Cu(In, Ga)Se$_2$ solar cells. In: Luque, A. & Hegedus, S. (eds.) *Handbook of Photovoltaic Science and Engineering*. Wiley: Hoboken, NJ, pp. 567–611.

Shah, A. (2018) Thin-film silicon solar cells. In: Kalogirou, S.A. (ed.) *McEvoy's Handbook of Photovoltaics: Fundamentals and Applications*, 3rd edition. Academic Press: Cambridge, MA, pp. 235–299.

Shen, X., Yang, M., Zhang, C., Qiao, Z., Wang, H. & Tang, C. (2018) Utilizing magnetron sputtered AZO-ITO bilayer structure as transparent conducting oxide for improving the performance of flexible CIGS solar cell. *Superlattices and Microstructures* 123, 251–256. doi: 10.1016/j.spmi.2018.09.001.

Shockley, W. & Queisser, H.J. (1961) Detailed balance limit of efficiency of pn junction solar cells. *Journal of Applied Physics* 32(3), 510. doi: 10.1063/1.1736034.

Tang, Q., Shen, H., Yao, H., Gao, K., Jiang, Y. & Liu, Y. (2018) Dopant-free random inverted nanopyramid ultrathin c-Si solar cell via low work function metal modified ITO and TiO_2 electron transporting layer. *Journal of Alloys and Compounds* 769, 951–960. doi: 10.1016/j.jallcom.2018.08.072.

Turala, A., Jaouad, A., Masson, D.P., Fafard, S., Arès, R. & Aimez, V. (2013) Isolation of III-V/Ge multijunction solar cells by Wet Etching. *International Journal of Photoenergy* 2013, 583867. doi: 10.1155/2013/583867.

Venkata Krishnaiah, K., Venkatalakshmamma, P., Basavapoornima, C., Martín, I.R., Soler-Carracedo, K., Hernández-Rodríguez, M.A., Venkatramu, V. & Jayasankar, C.K. (2017) Er3+-doped tellurite glasses for enhancing a solar cell photocurrent through photon upconversion upon 1500nm citation. *Materials Chemistry and Physics* 199, 67–72. doi: 10.1016/j.matchemphys.2017.06.003.

Vidal Lorbada, R., Walter, T., Fuertes Marrón, D., Lavrenko, T. & Muecke, D. (2019) A deep insight into the electronic properties of CIGS modules with monolithic interconnects based on 2D simulations with TCAD. *Coatings* 9(2), 128. doi: 10.3390/coatings9020128.

Yamaguchi, M. (2020) High-efficiency GaAs-based solar cells. In: Rahman, M.M. et al. (eds.) *Post-Transition Metals*. IntechOpen: London. doi:10.5772/intechopen.94365.

Yao, H. & Tang, Q. (2020) Luminescent anti-reflection coatings based on down-conversion emission of Tb^{3+}-Yb^{3+} co-doped $NaYF_4$ nanoparticles for silicon solar cells applications. *Solar Energy* 211, 446–452. doi: 10.1016/j.solener.2020.09.084.

Zaidi, B. (2018) Introductory chapter: introduction to photovoltaic effect. In: Zaidi, B. (ed.) *Solar Panels and Photovoltaic Materials*. IntechOpen: London. doi: 10.5772/intechopen.74389.

Zdyb, A., Krawczak, E. & Gułkowski, S. (2018) The influence of annealing on the properties of ZnO:Al layers obtained by RF magnetron sputtering. *Opto-Electronics Review* 26(3), 247–251. doi: 10.1016/j.opelre.2018.07.002.

Zhou, X., Deng, Y., Jiang, S., Xiang, G., Li, L., Tang, X., Luo, X., Pang, Y. & Huang, Y. (2019) Investigation of energy transfer in Pr^{3+}, Yb^{3+} co-doped phosphate phosphor: the role of 3P0 and 1D2. *Journal of Luminescence* 209, 45–51. doi: 10.1016/j.jlumin.2019.01.010.

Zhu, L., Luo, G., Yin, X., Tan, B., Guo, X., Li, W., Zhang, J., Wu, L., Zeng, G. & Wang, W. (2020) Ohmic contact formation and activation for reduced graphene oxide and Sb bilayers contacted CdTe solar cells. *Materials Science in Semiconductor Processing* 118, 105186. doi: 10.1016/j.mssp.2020.105186.

Chapter 2
Organic Solar Cells

2.1 INTRODUCTION

One of the most promising photovoltaic technologies of the third generation are organic solar cells, which are inexpensive, lightweight, simply processable and compatible with flexible substrates. The progress made in the development of this technology, based on the introduction of newly developed materials and optimization of the organic cell structure, has led to significant improvements in their efficiency.

The semiconductor heterojunction constituting the essential part of the organic photovoltaic (OPV) cell consists of donor material donating electrons and transporting holes and acceptor material withdrawing electrons and transporting electrons. Planar junctions or bulk mixed junctions including binary and ternary blends can be employed in the cells of normal or inverted structure.

In this chapter, organic semiconductors are juxtaposed with inorganic semiconductors typically used in the photocells of the first generation and different architecture variants of OPV devices are described. The exemplary photocells based on widely utilized fullerene acceptors are presented. Due to the relatively high price, poor light harvesting and small bandgap, fullerene acceptors are substituted with non-fullerene materials that demonstrate tunable bandgap enabling good match to solar spectrum. The introduction of non-fullerene acceptors accompanied by modifications of structure and other materials used resulted in an efficiency exceeding 18% for a laboratory OPV cell. This chapter also presents the efforts made to improve light trapping by the photocells and to fix stability issues, which are crucial in the context of future applications and commercialization of the OPV technology.

2.2 ORGANIC SEMICONDUCTORS

Organic semiconductors are solid-state materials, either polymers or small molecules (Rodriquez et al., 2016), in which high level of conjugation occurs. In conjugated molecules, a series of single and double bonds are arranged along the molecule and the electrons associated with the double bonds formed by π molecular orbital are less tightly localized, which enables them to participate in conduction of charge. Organic semiconductors form molecular crystalline materials or amorphous layers based on van der Waals interactions, unlike inorganic semiconductors, where chemical covalent bonds dominate. In organic semiconductors, the highest occupied molecular orbital

DOI: 10.1201/9781003196785-2

Table 2.1 The comparison of inorganic and organic semiconductor properties, based on Forrest (2015) and Dyer-Smith et al. (2018)

Property	Inorganic semiconductor	Organic semiconductor
Type of bonds	Covalent/ionic	van der Waals
Energy levels	Energy bands	Molecular orbitals
Absorption coefficient (cm^{-1})	10^3–10^4	10^5–10^6
Dielectric constant	10–15	2–5
Exciton type	Mott-Wannier	Frenkel
Binding energy of exciton (eV)	0.01–0.1	0.3–0.5
Exciton size	Around 10 nm	Around 1 nm
Carrier mobility ($cm^2/V/s$)	1500	1–3
Exciton diffusion length (nm)	7–100	5–20
Charge transport	Transfer within energy bands	Hopping of exciton between localized sites
Application	Wafers or thin films	Thin films

(HOMO) and the lowest unoccupied molecular orbital (LUMO) energy levels can be distinguished and the difference between them is the equivalent to the energy bandgap in inorganics. The charge is transferred between organic molecules, mainly by a hopping mechanism, and the conductivity is lower than in inorganic semiconductors. Organic semiconductors can serve as a photoactive material in solar cells, as they have an absorption coefficient of $10^5 cm^{-1}$; however, the absorption bands they exhibit are rather narrow (Dyer-Smith et al., 2018). When the energy of photons of impinging light is equal to or higher than the HOMO and LUMO difference, the light energy can be absorbed, which results in the generation of electron-hole pairs. The exciton created in this way in organic semiconductors is a strongly localized quasiparticle composed of an electron and a hole, which has a high binding energy (0.3–0.5 eV) (Dyer-Smith et al., 2018). The generated quasiparticle is a small-radius Frenkel exciton that differs from Mott-Wannier excitons, which are delocalized, have the large size and occur in inorganic materials. The specific properties and comparison of inorganic and organic semiconductors are presented in Table 2.1. The tightly bound Frenkel excitons can move by hopping, which results in the decrease in carrier mobility. If carrier mobility is low, an active layer thickness has to be reduced to around 100 nm in order to limit recombination and allow the carriers to reach the electric contacts of the photocell. In thinner layers, the resistance decreases; however, simultaneously the absorption process is weakened. The separation of free charge carriers is difficult and cannot be realized only thermally because of the low dielectric constant of the material (about 2–4) and large binding energy of the exciton. Therefore, the dissociation can take place only if the exciton reaches the interface of electron-donating (donor) and electron-accepting (acceptor) organic layers.

2.3 DEVICE ARCHITECTURE AND PERFORMANCE

In an organic photovoltaic cell, the light is absorbed in an active layer composed of electron donor and acceptor molecules in which excitons are created; however, most of the light is usually absorbed by the donor material. The successive stages of the working mechanism of the cell are presented in Figure 2.1.

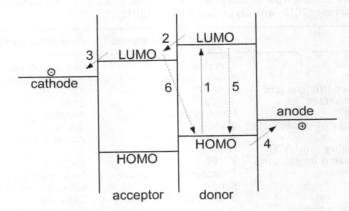

Figure 2.1 The scheme of donor and acceptor energy levels and organic cell operation.

The absorption of photon results in the creation of an exciton, which consists of an electron in the LUMO and a hole in the HOMO. The exciton diffuses to the interface of heterojunction, where it can be dissociated, which means that the electron can be transferred down the energy axis, to the LUMO of acceptor and then again to the lower energy level of the appropriate electrode. Simultaneously, the hole travels also down its energy axis, to the respective electrode on the other side of the cell. In the presented path of charge carriers, recombination can also take place, which is a deleterious process. The recombination of the electron-hole pair can happen before the splitting of exciton or after the splitting, across the interface. Energy loss into heat can also occur in the thermalization process, if the energy of absorbed photons exceeds the bandgap.

2.3.1 Single-layer organic photocell

In the simplest form of organic solar cell, a single layer of organic electronic material is placed between two electrodes. Upon illumination of the top transparent electrode, an exciton can be created within the organic semiconductor and then split into an electron and a hole in the process driven by the difference in work functions of electrode materials. If the upper electrode – anode (e.g., ITO or metal) – has a greater work function than the back metallic electrode – cathode (e.g., Ca, Al or Mg) – a potential difference occurs that helps to separate the two types of charge carriers, which is essential for the photovoltaic effect to occur.

2.3.2 Binary organic photocell

In a more complex structure of the OPV cell, two materials of different electron affinity and ionization energy are applied: donor that donates electrons and transports holes, and acceptor that withdraws electrons from the donor. The acceptor layer features a higher electron affinity and ionization potential than the donor. The excitons photogenerated in the active area of the cell diffuse due to the gradient of

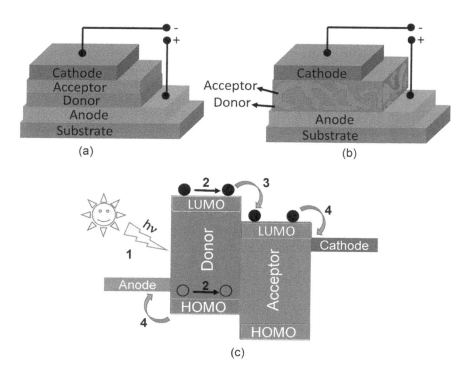

Figure 2.2 Structure of bilayer (a) and bulk heterojunction (b) organic photocell. Device working principle (c). (Copied under Creative Commons Attribution license from Kumaresan et al. (2014).)

concentration and then can be dissociated once they reach the interface between the two types of materials. For efficient charge separation and transport, the difference between the LUMO level of the acceptor material and the HOMO level of the donor material should be high enough. The architecture of a bilayer OPV cell, positions of energy levels and the path of charge carriers are shown in Figure 2.2.

The obtained open-circuit voltage depends mainly on the energy difference between the donor HOMO ($E_{d\,HOMO}$) and the acceptor LUMO ($E_{a\,LUMO}$). Voltage loss can be mainly ascribed to unfavorable HOMO energy level offset between the donor and acceptor compounds; however, the comprehensive insight shows numerous factors and mechanisms affecting the open-circuit voltage value, such as temperature, light intensity, recombination and carrier density (Elumalai & Uddin, 2016).

The thickness of the active layer is an essential parameter for the performance of the organic solar cell, since the trade-off has to be reached between the diffusion length, which tends to be low, and light absorption intensity, which achieves a sufficiently high value only in thicker layers.

2.3.3 Bulk heterojunction organic photocell

The improved performance can be achieved if bulk heterojunction (BHJ) is formed by a nanoscale photoactive blend of acceptor and donor materials and introduced in an organic photovoltaic cell. The BHJ photocell was reported for the first time in 1995 (Yu et al., 1995). Mixed materials with a network of donor and acceptor domains on the order of nanometers offer large surface of the interface, which may be more easily accessible for short-lifetime excitons. Figure 2.2b depicts the structure of an OPV cell with donor-acceptor mixture. However, the size of the domains forming the heterojunction cannot be too small to allow for effective charge transport and to ensure a good contact of acceptor material to electron-transporting electrode and donor material to hole-transporting electrode. The application of bulk heterojunction is beneficial, since it prevents the thickness limitations caused by low diffusion length of the excitons and allows using thicker layers that can absorb light more effectively.

2.3.3.1 Fullerene acceptors

The most commonly adopted approach is the application of BHJ composed of conjugated polymer as the donor and a fullerene derivative as the acceptor (Scharber et al., 2006). Fullerene derivatives, e.g., $PC_{60}BM$ (phenyl-C61-butyric-acid-methyl-ester) and $PC_{70}BM$ (phenyl-C71-butyric-acid-methyl-ester), which ensure good charge transport properties and spherical geometry, promoting the necessary separation of donor and acceptor, have been serving in organic cells for many years and still remain widely used materials.

Among the blends with fullerenes, the popular P3HT (poly(3-hexylthiophene-2,5-diyl) polymer is often used as the donor; however, the efficiencies of the cells based on P3HT are below 10%. At early stage, the mixture of P3HT and $C_{60}(CH_2)_2$ fullerene acceptor provided the power conversion efficiency (PCE) of 6.43% (He et al., 2014). The P3HT with fullerene bisadduct acceptor $IC_{70}BA$ delivered an efficiency of up to 7.4% and an improved open-circuit voltage of 0.87 V (Guo et al., 2012). Then, the improved PCE of 8.11% was obtained due to the introduction of isomer-free fullerene bisadducts (Xiao et al., 2016). Fullerenes functionalized with methoxy, ether and ester groups mixed with P3HT presented optimized morphology and an increase in the open-circuit voltage (Matsumoto et al., 2016). The further research focused on the determination of HOMO and LUMO positions indicated $IC_{60}TA$ and $IC_{70}TA$ as the promising acceptor materials for organic cells (Ruff et al., 2018). Recently, P3HT or PCDTBT (a copolymer composed of alternating thiophene–benzothiadiazole–thiophene and carbazole units) donors have been applied with monocyclopropanated fullerene derivatives, of which molecular structures are depicted in Figure 2.3. Among the presented fullerene derivatives exhibiting relatively low electron affinity, F1 donor blended with PCDTBT provided a PCE of 5% (Mumyatov et al., 2020). Figure 2.4 presents molecular structures of conjugated polymers and the architecture of the organic BHJ photocell. Another mix of materials: $PT_{12}OH$ donor and $PC_{61}BM$ fullerene acceptor, demonstrated a PCE of 4.83% and a remarkable stability in time (Lanzi & Pierini, 2021). The valuable alternative to $PC_{61}BM$ is the novel acceptor consisting of the planar molecule of rylene dye integrated with fullerene, which revealed an efficiency of over 8% (Feng et al., 2020). This achievement indicates a new interesting path in the research on fullerene acceptors.

Higher power conversion efficiencies of over 10% were demonstrated by the OPV devices based on fullerene derivatives applied in ternary blends, which will be presented in Section 2.3.5 (Ternary blends).

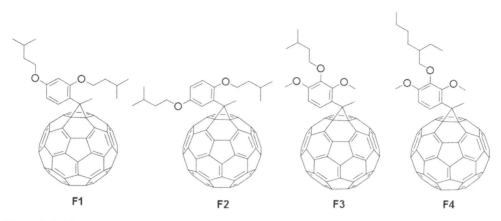

Figure 2.3 Monocyclopropanated fullerene derivatives – molecular structures. (Reprinted from Mumyatov et al. (2020). Copyright with permission from Elsevier.)

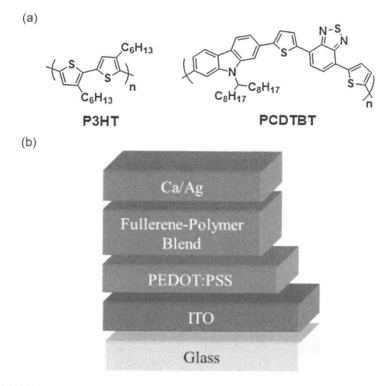

Figure 2.4 (a) Molecular structure of conjugated polymers; (b) the architecture of the OPV cell. (Reprinted from Mumyatov et al. (2020). Copyright with permission from Elsevier.)

2.3.3.2 Non-fullerene acceptors

Numerous disadvantages of fullerene acceptors such as limited absorption of visible light, problems with functionalization, degradation and high costs have motivated the search for non-fullerene (NF) acceptors (Duan et al., 2019; Hao et al., 2019; Suman & Prakash Singh, 2019; Wadsworth et al., 2019; Zhang et al., 2020).

Tunable absorption spectrum is an especially beneficial property of non-fullerene materials, which enables a good match of the bandgap value and solar spectrum. The commonly used organic semiconductors have a bandgap >1.1 eV exhibited by silicon, which is the most popular photovoltaic material. An example of fullerene-free acceptor with a bandgap of 1.57 eV is ITIC, which provides absorption from visible light (Vis) to near-infrared (NIR) as well as good electron mobility leading to the OPV device efficiency of 11.3% (Zheng et al., 2017).

However, many low-bandgap non-fullerene acceptors also have shown the potential in applications. To provide a broad range of light absorption when a low-bandgap material is used in the OPV device, donor-acceptor pairs can be matched so that their absorption bands complement each other. This strategy was applied in the cells with the following exemplary pairs of a larger-bandgap donor and a smaller-bandgap acceptor: J61 donor and ITIC acceptor provided a PCE of 9.53% (Bin et al., 2016), PffT2-FTAZ-2DT and IEIC showed a PCE of 7.3% (Lin et al., 2015), and PBDTS-DTBTO and ITIC led to a PCE of 9.09% (Yao et al., 2016). The PBDB-T donor with newly designed and synthetized non-fullerene acceptor INPIC-4F with E_g = 1.39 eV exhibited a PCE of 13.13% (Sun et al., 2018a). The application of ultra-narrow bandgap acceptor IECO-4F (E_g = 1.03 eV) mixed with PTB7-Th (E_g = 1.58 eV) also revealed a good PCE of 11% (Wang et al., 2018).

The research on low-bandgap acceptors led to the development of the highly conjugated organic semiconductor Y6 (BTP-4F) (E_g = 1.33 eV) which, when used together with PM6 donor (E_g = 1.81 eV), harvested a broad range of solar light and delivered an efficiency of 15.7%. Such an excellent result was confirmed both in the conventional structure of the ITO/PEDOT:PSS/PM6:Y6/PDINO/Al cell and in the inverted configuration ITO/ZnO/PM6:Y6/MoO$_3$/Ag with 100–150 nm thickness of the blend film (Yuan et al., 2019). Further investigations and proper selection of donors cooperating with the low-bandgap Y6 acceptor significantly boosted the efficiency. Two examples are L1 and D16 copolymer donors; they are used in the OPV cells with Ag or Al electrodes and provide satisfactory results. The molecular structures of L1 and D16 donors and Y6 acceptor, their absorption spectra and energy level diagram as well as current density vs. voltage characteristics of the photocells are shown in Figure 2.5. Thiolactone copolymer donor D16 when blended with Y6 demonstrated the highest PCE of 16.72%, V_{oc} of 0.85 V, J_{sc} of 26.61 mA/cm^2 and FF of 73.8% (Figure 2.5e) (Xiong et al., 2019). Further search and modifications of donors led to the usage of D18 donor with higher hole mobility and wide bandgap of 1.98 eV, which additionally improved the PCE of the OPV by up to 18.22% (Liu et al., 2020). Meanwhile, the research that focused on the structure of highly efficient Y6 resulted in the preparation of BTP-eC9 by the modification of Y6 in a few steps. This new acceptor characterized by good charge mobility and current density provided a PCE of 17.8% (Cui et al., 2020). Achieving such an outstanding performance of organic cells with Y6 was possible owing to the boost in short-circuit current and limitation of non-radiative energy loss.

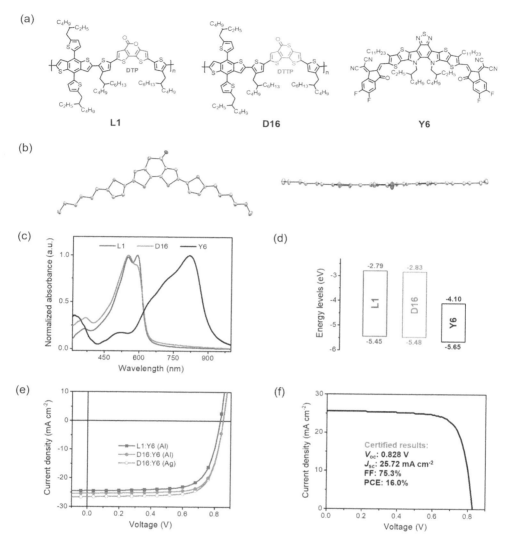

Figure 2.5 (a) The chemical structures of L1 and D16 donors and Y6 acceptor. (b) Top view (left) and side view (right) of the single-crystal structure of DTTP-2T with thiolactone unit DTTP occurring in D16. (c) Absorption spectra of L1, D16 and Y6 films. (d) Energy level diagram. (e) Current density vs. voltage curves for the solar cells based on three different donor-acceptor mixtures with Al or Ag electrode. (f) Current density vs. voltage characteristic for D16:Y6 solar cells with Ag electrode, which provided a certified PCE of 16%. (Reprinted from Xiong et al. (2019). Copyright with permission from Elsevier.)

In organic cells, not only the bandgap width of the active materials matters, proper selection and matching of donor and acceptor materials also decides the mutual arrangement of HOMO and LUMO levels. The difference between the donor HOMO and the acceptor LUMO determines the driving force necessary to split the exciton, as well as the open-circuit voltage of the OPV device. Efficient separation of charge at the donor-acceptor interface requires a driving force as high as 0.3–0.6 eV (Janssen & Nelson, 2013; Veldman et al., 2009); however, the use of novel non-fullerene acceptors enables exciton dissociation with a lower driving force, which is an advantage in terms of achieving higher V_{OC} and efficiency of the cell (Liu et al., 2016). The suitable positions of HOMO and LUMO should lead to a decrease in the energy loss without sacrificing the efficiency of exciton dissociation into free electron and hole.

Another crucial factor that directly influences the performance of a bulk heterojunction OPV device is the morphology of the donor-acceptor blend. In contrast to their fullerene counterparts, the non-fullerene materials exhibit a planar structure, which favors the miscibility of a donor and an acceptor and does not provide a sufficient degree of separation of the two phases. A good solution, which helps to maintain the separation of the phases, is the enhancement of the crystallinity of two active materials; however, crystals cannot be too large to avoid the reduction in the interface area. In the blend of the D18 donor and Y6 acceptor, which provided the high efficiency of organic cell (18.22%), aggregated nanofibers with a diameter of 20 nm were observed (Liu et al., 2020).

The morphology optimization and enhancement of performance of BHJ cells is also influenced by the presence of H-bonding in some organic semiconducting materials (Xiao et al., 2018). In donor materials, H-bonding is responsible for the self-assembly of small molecules, which is an important property related to better crystallinity and higher carrier mobility (Sun et al., 2015). In optimization of the morphology of the blend of the acceptor with a newly designed polymer donor (Gopalan et al., 2014), the H-bonding influences the interaction of donor and acceptor materials (interaction between polymer and fullerene derivatives). In this case, micelle-like structures of the polymer can be formed, which results in the improvement of performance (Li et al., 2015).

Another promising way of manipulating the morphology of the blend is the use of liquid crystalline materials. The application of novel small-molecule donors characterized by liquid crystalline properties can lead to the fabrication of high-performance OPV cells (Zhou et al., 2018; Zhang et al., 2017). For instance, the newly developed donor BTR-Cl, with properly ordered morphology and an active layer of 110 nm thickness, resulted in an average PCE of 13.29% (Chen et al., 2019a).

2.3.4 Inverted structure of organic photocell

The organic photocells in which a high work function ITO is employed as anode and a low work function metal is employed as cathode are the cells of the conventional structure. However, much interest is attracted by inverted organic cells, which differ in the sequence of the layers. The inverted photocell consists of a cathode, which is usually the ITO covered by ZnO electron transport layer (ETL), the layer of the photoactive blend and then the hole transport layer (HTL) (e.g., MoO_3), and the Ag anode. The example is an inverted organic photovoltaic cell of the following structure: ITO/ZnO/J71:NFA/MoO_3/Ag, which delivered a PCE of 9.61% with the use of non-fullerene DTC(4R)-IC and 10.89% when DTC(4R)-4FIC obtained upon fluorination was introduced into the device (Chen et al., 2019b).

The inverted photocells are known for better stability and lifetime, as well as broadband light harvesting. The additional advantage of the OPV inverted structure is the elimination of the PEDOT:PSS/ITO interface where degradation and instability issues occur.

In the cell with an inverted structure, the ITO coverage of cathode is modified to achieve a better contact with the photoactive layer by the coating of ZnO or TiO_2 serving for electron transport. The most widely used electron transport layer is ZnO, characterized by good conductivity and transparency, although the defects occurring in the ZnO structure are the reason for charge trapping. Therefore, many studies aimed to improve the ZnO performance by surface defects passivation through the introduction of solution-processed quantum dots or modification of ZnO surface. The beneficial role of quantum dots passivating ZnO surface defects was demonstrated in the inverted OPV cell based on P3HT/PC_{61}BM, which delivered 3.83% PCE. In such a device, electron extraction was promoted by ZnO enriched with 10% of MoS_2 quantum dots and the suppression of charge recombination enhanced the efficiency compared to a photocell based on pristine ZnO (Xie et al., 2019). The modification of the interface between ZnO and PTB7-Th/PC_{71}BM photoactive layer by the incorporation of a commercial dispersant product provided a PCE of 9.32% (Yu et al., 2019a). In this case, the improvement of performance in comparison with the control device was obtained due to the reduction in the surface roughness and the enhancement of electron mobility.

The substitution of ZnO by other ETL buffer layer materials was also tested; for example, the fullerene derivative DMAPA-C_{60} provided an inverted photocell efficiency of 7.43% PCE due to better contact with ITO (Wang et al., 2017) and the deposition of a smooth layer of bathocuproine (BCP) doped by fullerene C_{70} on ITO led to a PCE of 3.28% (Jafari et al., 2019). Figure 2.6 shows the structure of the inverted OPV cell with BCP:C_{70} in the role of electron transport layer and exciton blocking layer, as well as the energy level diagram of the device. The further enhancement of the inverted photocell PCE to over 11% was achieved by replacing the ZnO by a planar coronene derivative as ETL (Yu et al., 2017).

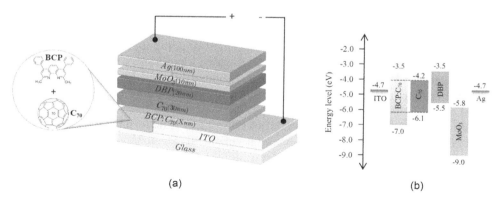

Figure 2.6 (a) Inverted OPV cell with BCP:C_{70} as electron transport layer and exciton blocking layer. (b) Energy level diagram of the OPV device. (Copied with no changes under Creative Commons Attribution 4.0 International License from Jafari et al. (2019).)

The anode, which is the top electrode collecting holes in an inverted OPV device, is usually covered by a high work function metal such as Ag, or Au with PEDOT:PSS, or MoO$_3$ modified Ag (Zhang et al., 2010, 2011). The application of PEDOT:PSS conductive polymer as a hole transport layer is possible for its high work function (Lee et al., 2020) and even beneficial due to relatively low price and flexibility. This approach was realized in all-solution-processed cells where PEDOT:PSS was used to improve conductivity and transparency, which led to 1.81% PCE (Kim et al., 2020a).

2.3.5 Ternary blends

Due to the pivotal role of the composition of photoactive blend, a huge effort has recently been devoted to employing different modifications, including the enrichment of the blend with a third additional ingredient. This strategy is intended to expand the spectral range of light absorption by complementing the absorption bands exhibited by all three components. By adding the third component, the formation of cascade energy levels, improvement of charge and energy transfer and beneficial effect on blend morphology can be obtained. The approach with either two donors and one acceptor or two acceptors and one donor is possible. Complementary properties of the blend components have revealed their advantages in numerous studies (Bi & Hao, 2019; Weng et al., 2019).

Among fullerene derivatives, the PC$_{71}$BM acceptor combined with the PBDTTPD-HT donor and DRCN$_5$T as a secondary donor provided an enhanced overall performance owing to properly adjusted energy levels and cascading charge transfer between them (Kim et al., 2020b). The investigation of the cell with the blend consisting of two polymer donors (wide bandgap F8T2, E_g=2.4 eV, and narrow bandgap PTB7, E_g=1.6 eV) mixed with fullerene acceptor PC$_{61}$BM resulted in a performance enhancement of 25% over the binary control device (Farinhas et al., 2017). The same dual-polymer donor mixture with PC$_{71}$BM as an acceptor led to similar effects. The recent achievement of 16.5% PCE was possible by the application of PBDB-TF donor and Y6 acceptor blend with the addition of PC$_{61}$BM as a second acceptor, which enhanced the electronic mobility and finally the overall performance of the cell (Yu et al., 2019b). Another solution, in which the same pair of materials, i.e., PM6 donor and Y6 acceptor, was enriched by an additional non-fullerene MF1 acceptor, delivered an efficiency of over 17.2%. The absorption ranges of components of the mixture are complimentary, as it is visible in Figure 2.7, which also depicts the structure of the photocell and energy levels scheme. The good compatibility of Y6 and MF1 proven by experimental methods, e.g., Raman mapping and cyclic voltammetry, suggests formation of an alloy-like state (An et al., 2020). Replacing MF1 by another non-fullerene acceptor has recently provided even better efficiency of the ternary organic cell. The blend composed of PM6:Y6 complemented with C8-DTC achieved a PCE of 17.52% due to confirmed better hole and electron mobility compared to the binary PM6:Y6 system, and the desired nanosized network morphology, which brings efficient charge separation and charge transport (Ma et al., 2020; Chang et al., 2021).

Among the materials used in ternary organic cells, there were also non-fullerene small-molecule acceptors IT-2Cl and IT-4Cl, in which C–Cl bonds have a large dipole moment that enhances charge transfer and broadens the absorption spectrum. When

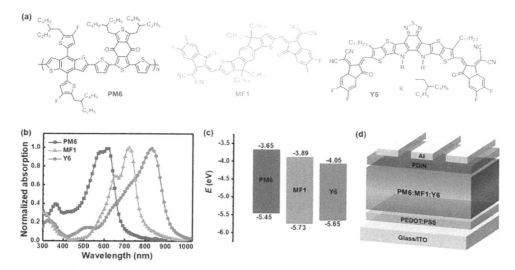

Figure 2.7 (a) Molecular structures of PM6 donor, Y6 acceptor and MF1 acceptor. (b) Normalized absorption spectra of PM6, MF1 and Y6 films. (c) Energy levels of PM6, MF1 and Y6. (d) Schematic structure of a ternary polymer solar cell. (Reprinted from An et al. (2020). Copyright with permission from Elsevier.)

these two acceptors were incorporated in the polymer donor blend, the efficiency of 14% and FF of 76% were achieved (Zhang et al., 2018). Other studies showed that the addition of non-fullerene acceptor, IDFBR, into a PffBT4T-2OD:PC$_{71}$BM blend resulted in the formation of cascading energy levels alignment and, in consequence, improved charge transport (Xu et al., 2018). Small molecules were also used as donors in ternary blends, leading to an efficiency over 6.5% (Huo et al., 2018).

In general, the losses of energy occurring in most ternary systems are the result of the unsatisfactory fit of the donor-acceptor energy levels and the complex relation between short-circuit current and open-circuit voltage of the cell. In search of the optimum solution to these problems, the ternary systems of inverted structure still attract the attention. For instance, the simultaneous improvement of I_{SC} and V_{OC} owing to enhanced absorption in the range of 600–750 nm and adjusted energy levels cascade was achieved by ternary cells with the inverted structure of ITO/ZnO/PTB$_7$-Th:IT-M:PC$_{71}$BM/MoO$_x$/Ag, the PCE of which reached up to 9.87% and FF to 67.5% (Sun et al., 2018b).

2.4 LIGHT TRAPPING

In organic cells, the optical losses occur mainly due to reflection of light from the electrodes and incomplete absorption as a consequence of partial transparency of the active layer. Therefore, a key challenge is to develop the light trapping methods that are effective in thin layers used in OPV devices.

One of the methods of avoiding reflection losses is the fabrication of V-shaped configuration of adjacent photocells, which can be relatively easily achieved by folding the structures obtained on a flexible substrate. The light can be reflected from the side surfaces, which increases the chances of light absorption. Additional improvement in performance triggered by absorption broadening is obtained if two different active materials exhibiting various bandgaps are placed on the two sides of the V-shaped device (Figure 2.8) (Tvingstedt et al., 2007). The fold angle affects the overall absorption of adjacent photocells and, in consequence, the J-V characteristic of the cells, which is visible in Figure 2.9.

In some cases, the folding of the cell may result in the formation of defects; thus, the application of microprism substrate covered with a homogenous active material is a preferred solution since in such a structure a second absorption process can occur (Niggemann et al., 2008).

The modifications aiming at trapping of light also include a back metal electrode, which can be structured or covered by randomly scattering structures (e.g., a layer of white scattering material, such as $BaSO_4$ or TiO_2) in order to enhance light absorption and, in consequence, the photocurrent. The beneficial method is the embedding of dielectric scatters such as TiO_2 in the form of nanoparticles (Tang et al., 2014).

Trapping of light can be also enhanced by the assistance of plasmons in metal nanoparticles incorporated in the structure of the organic cell. Metal nanoparticles

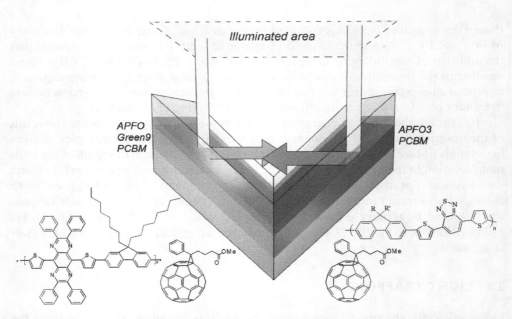

Figure 2.8 The architecture of a folded tandem OPV cell. The molecular structures of a high-bandgap $APFO_3$ polymer with absorption up to 650nm, a low-bandgap APFO-Green9 polymer with absorption extended up to 900nm and PCBM acceptor used in the cell are also shown. (Reprinted from Tvingstedt et al. (2007), with the permission of AIP Publishing.)

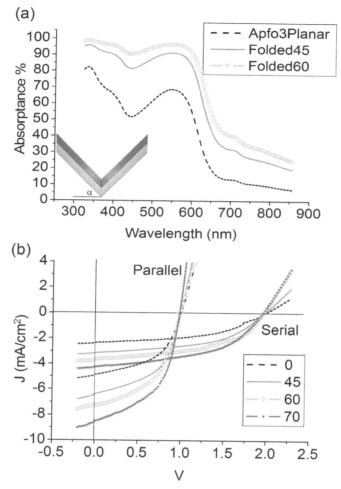

Figure 2.9 (a) The absorptance of a photocell with APFO$_3$ polymer on both sides of V-shaped structure. (b) J-V characteristics of parallel and series connections of folded cells for different fold angles (0°–70°) measured under AM1.5 100 mW/cm^2 solar light. (Reprinted from Tvingstedt et al. (2007), with the permission of AIP Publishing.)

introduced in the top layer of the cell limit the reflection of incident light by scattering the light into the layer of high refractive index, which increases the optical path length (Figure 2.10a). Under illumination, the excitation of plasmons results in localized surface plasmon resonance, which enhances the photoabsorption in the range over the resonant wavelength. This effect can be tuned by adjusting suitable size, concentration and location of the applied metal nanoparticles. The best results can be achieved when the metal nanoparticles are placed between an active layer and back electrode or blended with an active layer (Figure 2.10b and c) (Lim et al., 2018).

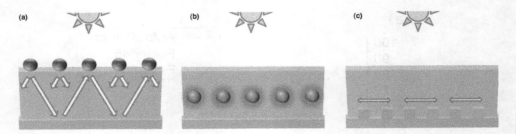

Figure 2.10 (a) Metal nanoparticles placed in the top layer of the OPV cell scatter light into the substrate. (b) Excited localized surface plasmons of metal nanoparticles placed in the photoactive layer increase the absorption. (c) Surface plasmon polaritons at the back contact of the photocell. (Reprinted from Tang et al. (2014), with permission from Elsevier.)

The nanoparticles of rare earth also take part in the enhancement of light absorption in OPV devices through the upconversion process. To this aim, the rare earth nanoparticles have to be embedded at the interface of the photoactive material with transparent electrode or the back electrode or within the layer of the photoactive material. The upconversion materials enable harvesting low-energy photons that are not absorbed by the semiconductor and then emit photons of higher energy, which are useful in photovoltaic conversion (Kesavan et al., 2019).

2.5 STABILITY

From the commercialization point of view, the durability of the organic solar cells and the maintenance of the performance is an important issue that should be addressed at the stage of laboratory investigations (Peters et al., 2011). The overview and analysis of degradation mechanisms is especially important in the case of organic photovoltaics, since the cells based on organic materials are more likely to degrade than the silicon or thin-film solar devices of first or second generations.

Extrinsic degradation occurs when the OPV cell works under ambient external conditions where moisture or oxygen enters the device and reacts with the components of the cell. The electrodes made of low work function metals (e.g., Al and Ca) become oxidized and, in consequence, change the work function, under the influence of oxygen and water, in the process that is independent of illumination. Chemical analysis indicates that polymers also react with oxygen, which leads to bleaching of the absorber layer. The investigations show that denser materials of crystalline structure are more resistant to photooxidation (Mateker & McGehee, 2017). The failure in encapsulation of the photocell results in rapid degradation; therefore, the tight sealing is crucial to avoid the delamination allowing the diffusion of water or oxygen into the structure of the cell. A variety of organic and inorganic compounds in the form of single layer or multilayer are applied as encapsulating materials (Li et al., 2021).

Intrinsic degradation processes occur both in the dark and under illumination (Mateker et al., 2015). In the dark, the changes of blend morphology in BHJ OPV are observed as well as the diffusion of atoms from the electrode material (e.g., In and Al) into the absorber, which disturbs the operation of the device (Voroshazi et al., 2011). The initial exposition of organic cell to sunlight can lead to rapid degradation that decreases the efficiency of new cells. Photo-induced reactions (e.g., formation of fullerene dimers) can be the reason for up to 10–50% losses in efficiency; therefore, the limitation of photochemical reactions by the elimination of impurities and improvement of crystallinity in active material are necessary.

The assessment of stability of the photocells can be performed under real external conditions or in accelerated tests in which concentrated light and elevated temperature are applied. The improvement of stability has recently been achieved by the cell based on the bulk heterojunction of $PC_{61}BM$ with new polythiophene derivative PT12OH. The beneficial performance over a period of 120 h was confirmed owing to the stabilized nanophase-separated morphology of the blend induced by –OH functionality. No additional thermal, chemical or photochemical treatment was applied in those investigations (Lanzi & Pierini, 2021). Ternary OPV devices, displaying a recent significant improvement in performance, suffer from the problems with the long-term stability of the morphology of the blend, which is a crucial issue influencing the performance. However, the stability of the ternary cell ($PIDTBT:PC_{71}BM:ITIC$) over 410 days was demonstrated. The persistence of the cells was investigated according to the procedure of Krebs, and the stable PCE value during 59 weeks of storage was proven. The procedure involves the evaluation of working parameters in dark, at room temperature, in nitrogen (Chen et al., 2018).

2.6 SUMMARY

A substantial improvement in the efficiency and lifetime has been achieved by OPV cells owing to the optimization of the absorber material, modification of architectures, introduction of alternative electrode materials and more effective encapsulation. A progress in the development of the OPV technology, as evidenced by the efficiency of the laboratory photocells exceeding 18% and achievements in their stability improvement, indicates the prospective possibility of spreading this technology in the photovoltaic market. Moreover, the special features of an OPV cell, such as lightweight and possible flexibility, favor a variety of applications.

REFERENCES

An, Q., Wang, J., Gao, W., Ma, X., Hu, Z., Gao, J., Xu, C., Hao, M., Zhang, X., Yang, C. & Zhang, F. (2020) Alloy-like ternary polymer solar cells with over 17.2% efficiency. *Science Bulletin* 65(7), 538–545. doi: 10.1016/j.scib.2020.01.012.

Bi, P. & Hao, X. (2019) Versatile ternary approach for novel organic solar cells: a review. *RRL Solar* 3(1), 1800263. doi: 10.1002/solr.201800263.

Bin, H., Zhang, Z.G., Gao, L., Chen, S., Zhong, L., Xue, L., Yang, C. & Li, Y. (2016) Non-fullerene polymer solar cells based on alkylthio and fluorine substituted 2D-conjugated polymers reach 9.5% efficiency. *Journal of American Chemical Society* 138, 4657–4664. doi: 10.1021/jacs.6b01744.

Chang, L., Sheng, M., Duan, L. & Uddin, A. (2021) Ternary organic solar cells based on non-fullerene acceptors: a review. *Organic Electronics* 90, 106063. doi: 10.1016/j.orgel.2021.106063.

Chen, C.-P., Li, Y.-C., Tsai, Y.-Y. & Lu, Y.-W. (2018) Efficient ternary polymer solar cells with a shelf-life stability for longer than 410 days. *Solar Energy Materials and Solar Cells* 183, 120–128. doi: 10.1016/j.solmat.2018.04.014.

Chen, H., Hu, D., Yang, Q., Gao, J., Fu, J., Yang, K., He, H., Chen, S., Kan, Z., Duan, T., Yang, C., Ouyang, J., Xiao, Z., Sun, K. & Lu, S.(2019a) All-small-molecule organic solar cells with an ordered liquid crystalline donor. *Joule* 3(12), 3034–3047. doi: 10.1016/j.joule.2019.09.009.

Chen, T.-W., Hsiao, Y.-T., Lin, Y.-W., Chang, C.-C., Chuang, W.-T., Li, Y. & Hsu, C.-S. (2019b) Fluorinated heptacyclic carbazole-based ladder-type acceptors with aliphatic side chains for efficient fullerene-free organic solar cells. *Materials Chemistry Frontiers* 3, 829–835. doi: 10.1039/C9QM00005D.

Cui, Y., Yao, H., Zhang, J., Xian, K., Zhang, T., Hong, L., Wang, Y., Xu, Y., Ma, K., An, C., He, C., Wei, Z., Gao, F. & Hou, J. (2020) Single-junction organic photovoltaic cells with approaching 18% efficiency. *Advanced Materials* 32(19), 1908205. doi: 10.1002/adma.201908205.

Duan, L., Elumalai, N.K., Zhang, Y. & Uddin, A. (2019) Progress in non-fullerene acceptor based organic solar cells. *Solar Energy Materials and Solar Cells* 193, 22–65. doi: 10.1016/j.solmat.2018.12.033.

Dyer-Smith, C., Nelson, J. & Li, Y. (2018) Organic solar cells. In: Kalogirou, S.A. (ed.) *McEvoy's Handbook of Photovoltaics: Fundamentals and Applications*, 3rd edition. Academic Press: Cambridge, MA, pp. 561–590.

Elumalai, N.K. & Uddin, A. (2016) Open circuit voltage of organic solar cells: an in-depth review. *Energy & Environmental Science* 9(2), 391–410. doi: 10.1039/C5EE02871J.

Farinhas, J., Oliveira, R., Hansson, R., Ericsson, L.K.E., Moons, E., Morgado, J. & Charas, A. (2017) Efficient ternary organic solar cells based on immiscible blends. *Organic Electronics* 41, 130–136. doi: 10.1016/j.orgel.2016.12.009.

Feng, J., Fu, H., Jiang, W., Zhang, A., Ryu, H.S., Woo, H.Y., Sun, Y. & Wang, Z. (2020) Fullerylenes: paving the way for promising acceptors. *ACS Applied Materials Interfaces* 12(26), 29513–29519. doi: 10.1021/acsami.0c05548.

Forrest, S.R. (2015) Excitons and the lifetime of organic semiconductor devices. *Philosophical Transactions: Series A, Mathematical, Physical, and Engineering Sciences* 373, 20140320. doi: 10.1098/rsta.2014.0320.

Gopalan, S.A., Seo, M.H., Anantha-Iyengar, G., Han, B., Lee, S.-W., Kwon, D.-H., Lee, S.-H. & Kang, S.-W. (2014) Mild wetting poor solvent induced hydrogen bonding interactions for improved performance in bulk heterojunction solar cells. *Journal of Materials Chemistry A* 2(7), 2174–2186. doi: 10.1039/C3TA13875E.

Guo, X., Cui, C., Zhang, M., Huo, L., Huang, Y., Hou, J. & Li, Y. (2012) High efficiency polymer solar cells based on poly(3-hexylthiophene)/indene-C_{70} bisadduct with solvent additive. *Energy & Environmental Science* 5, 7943–7949. doi: 10.1039/C2EE21481D.

Hao, M., Liu, T., Xiao, Y., Ma, L.-K., Zhang, G., Zhong, C., Chen, Z., Luo, Z., Lu, X., Yan, H., Wang, L. & Yang, C. (2019) Achieving balanced charge transport and favorable blend morphology in non-fullerene solar cells via acceptor end group modification. *Chemistry of Materials* 31(5), 1752–1760. doi: 10.1021/acs.chemmater.8b05327.

He, D., Du, X., Xiao, Z. & Ding, L. (2014) Methanofullerenes, $C_{60}(CH_2)_n$ (n=1, 2, 3), as building blocks for high-performance acceptors used in organic solar cells. *Organic Letters* 16(2), 612–615. doi: 10.1021/ol4035275.

Huo, Y., Zhu, J., Wang, X.-Z., Yan, C., Chai, Y.-F., Chen, Z.-Z., Zhan, X. & Zhang, H.-L. (2018) Small molecule donors based on benzodithiophene and diketopyrrolopyrrole compatible with both fullerene and non-fullerene acceptors. *Journal of Materials Chemistry C* 6, 5843–5848. doi: 10.1039/C8TC00799C.

Jafari, F., Patil, B.R., Mohtaram, F., Cauduro, A.L.F., Rubahn, H.-G., Behjat, A. & Madsen, M. (2019) Inverted organic solar cells with non-clustering bathocuproine (BCP) cathode interlayers obtained by fullerene doping. *Scientific Reports* 9, 10422. doi: 10.1038/s41598-019-46854-w.

Janssen, R.A.J. & Nelson, J. (2013) Factors limiting device efficiency in organic photovoltaics. *Advanced Materials* 25, 1847–1858. doi: 10.1002/adma.201202873.

Kesavan, A.V., Kumar, M.P., Rao, A.D., & Ramamurthy, P.C. (2019) Light management through up-conversion and scattering mechanism of rare earth nanoparticle in polymer photovoltaics. *Optical Materials* 94, 286–293. doi: 10.1016/j.optmat.2019.04.057.

Kim, D.H., Lee, D.J., Kim, B., Yun, C. & Kang, M.H. (2020a) Tailoring PEDOT:PSS polymer electrode for solution-processed inverted organic solar cells. *Solid-State Electronics* 169, 107808. doi: 10.1016/j.sse.2020.107808.

Kim, D.H., Ryu, J., Wibowo, F.T.A., Park, S.Y., Kim, J.Y., Jang, S.-Y., & Cho, S. (2020b) Elimination of charge transfer energy loss by introducing a small-molecule secondary donor into fullerene-based polymer solar cells. *ACS Applied Energy Materials* 3(9), 8375–8382. doi: 10.1021/acsaem.0c01000.

Kumaresan, P., Vegiraju, S., Ezhumalai, Y., Yau, S.L., Kim, C., Lee, W.-H., Chen, M.-C. (2014) Fused-thiophene based materials for organic photovoltaics and dye-sensitized solar cells. *Polymers* 6, 2645–2669. doi: 10.3390/polym6102645.

Lanzi, M. & Pierini, F. (2021) Efficient and thermally stable BHJ solar cells based on a soluble hydroxy-functionalized regioregular polydodecylthiophene. *Reactive and Functional Polymers* 158, 104803. doi: 10.1016/j.reactfunctpolym.2020.104803.

Lee, H., Park, S., Cho, I.-C., Ryu, M.-Y., Kang, D., Yi, Y. & Lee, H. (2020) Metal-electrode-free inverted organic photovoltaics using electrospray-deposited PEDOT:PSS and spin-coated HAT-CN exciton blocking layer. *Current Applied Physics* 20(2), 277–281. doi: 10.1016/j.cap.2019.11.013.

Li, F., Yager, K.G., Dawson, N.M., Jiang, Y.B., Malloy, K.J. & Qin, K.Y. (2015) Nano-structuring polymer/fullerene composites through the interplay of conjugated polymer crystallization, block copolymer self-assembly and complementary hydrogen bonding interactions. *Polymer Chemistry* 6(5), 721–731. doi: 10.1039/C4PY00934G.

Li, Y., Huang, X., Ding, K., Sheriff Jr, H.K.M., Ye, L., Liu, H., Li, C.-Z., Ade, H. & Forrest, S.R. (2021) Non-fullerene acceptor organic photovoltaics with intrinsic operational lifetimes over 30 years. *Nature Communication* 12, 5419. doi: 10.1038/s41467-021-25718-w.

Lim, S.J., Kim, D.U., Song, J.-H. & Yu, J.-W. (2018) Enhanced performance of semi-transparent OPV with nanoparticle reflectors. *Organic Electronics* 59, 314–318. doi: 10.1016/j.orgel.2018.05.054.

Lin, H., Chen, S., Li, Z., Lai, J.Y.L., Yang, G., McAfee, T., Jiang, K., Li, Y., Liu, Y., Hu, H., Zhao, J., Ma, W., Ade, H. & Yan, H. (2015) High-performance non-fullerene polymer solar cells based on a pair of donor–acceptor materials with complementary absorption properties. *Advanced Materials* 27(45), 7299–7304. doi: 10.1002/adma.201502775.

Liu, J., Chen, S., Qian, D., Gautam, B., Yang, G., Zhao, J., Bergqvist, J., Zhang, F., Ma, W., Ade, H., Inganäs, O., Gundogdu, K., Gao, F. & Yan, H. (2016) Fast charge separation in a non-fullerene organic solar cell with a small driving force. *Nature Energy* 1, 16089. doi: 10.1038/nenergy.2016.89.

Liu, Q., Jiang, Y., Jin, K., Qin, J., Xu, J., Li, W., Xiong, J., Liu, J., Xiao, Z., Sun, K., Yang, S., Zhang, X. & Ding, L. (2020) 18% efficiency organic solar cells. *Science Bulletin* 65(4), 272–275. doi: 10.1016/j.scib.2020.01.001.

Ma, Q., Jia, Z., Meng, L., Zhang, J., Zhang, H., Huang, W., Yuan, J., Gao, F., Wan, Y., Zhang, Z. & Li, Y. (2020) Promoting charge separation resulting in ternary organic solar cells efficiency over 17.5%. *Nano Energy* 78, 105272. doi: 10.1016/j.nanoen.2020.105272.

Mateker, W.R. & McGehee, M.D. (2017) Progress in understanding degradation mechanisms and improving stability in organic photovoltaics. *Advanced Materials* 29(10), 1603940. doi: 10.1002/adma.201603940.

Mateker, W.R., Sachs-Quintana, I.T., Burkhard, G.F., Cheacharoen, R. & McGehee, M.D. (2015) Minimal long-term intrinsic degradation observed in a polymer solar cell illuminated in an oxygen-free environment. *Chemistry of Materials* 27, 404–407. doi: 10.1021/cm504650a.

Matsumoto, F., Iwai, T., Moriwaki, K., Takao, Y., Ito, T., Mizuno, T. & Ohno, T. (2016) Controlling the polarity of fullerene derivatives to optimize nanomorphology in blend films. *ACS Applied Materials Interfaces* 8(7), 4803–4810. doi: 10.1021/acsami.5b11180.

Mumyatov, A.V., Goryachev, A.E., Prudnov, F.A., Mukhacheva, O.A., Sagdullina, D.K., Chernyak, A.V., Troyanov, S.I. & Troshin, P.A (2020) Monocyclopropanated fullerene derivatives with decreased electron affinity as promising electron acceptor materials for organic solar cells. *Synthetic Metals* 270, 116565. doi: 10.1016/j.synthmet.2020.116565.

Niggemann, M., Riede, M., Gombert, A. & Leo, K. (2008) Light trapping in organic solar cells. *Applications and Materials Science* 205(12), 2862–2874. doi: 10.1002/pssa.200880461.

Peters, C.H., Sachs-Quintana, I.T., Kastrop, J.P., Beaupre, S., Leclerc, M. & McGehee, M.D. (2011) High efficiency polymer solar cells with long operating lifetimes. *Advanced Energy Materials* 1(4), 491–494. doi: 10.1002/aenm.201100138.

Rodriquez, D., Savagatrup, S., Valle, E., Proctor, C.M., McDowell, C., Bazan, G.C., Nguyen, T.-Q. & Lipomi, D.J. (2016) Mechanical properties of solution-processed small-molecule semiconductor films. *ACS Applied Materials and Interfaces* 8(18), 11649–11657. doi: 10.1021/acsami.6b02603.

Ruff, A., Qian, X., Porfyrakis, K. & Ludwigs, S. (2018) Effect of the type and number of organic addends on fullerene acceptors for n-type electronic devices: redox properties and energy levels. *Chemistry Select* 3(21), 5778–5785. doi: 10.1002/slct.201800837.

Scharber, M.C., Mühlbacher, M.D, Koppe, M., Denk, P., Waldauf, C., Heeger, A.J. & Brabec, C.J.(2006) Design rules for donors in Bulk-Heterojunction solar cells: towards 10% energy-conversion efficiency. *Advanced Materials* 18(6), 789–794. doi: 10.1002/adma.200501717.

Suman, S. & Prakash Singh, S. (2019) Impact of end groups on the performance of non-fullerene acceptors for organic solar cell applications. *Journal of Materials Chemistry A* 7, 22701–22729. doi: 10.1039/C9TA08620J.

Sun, K., Xiao, Z., Lu, S., Zajaczkowski, W., Pisula, W., Hanssen, F., White, J.M., Williamson, R.M., Subbiah, J., Ouyang, J., Holmes, A.B., Wong, W.W.H. & Jones, D.J. (2015) A molecular nematic liquid crystalline material for high-performance organic photovoltaics. *Nature Communications* 6, 6013. https://www.nature.com/articles/ncomms7013.

Sun, J., Ma, X., Zhang, Z., Yu, J., Zhou, J., Yin, X., Yang, L., Geng, R., Zhu, R., Zhang, F. & Tang, W. (2018a) Dithieno[3,2-b:2′, 3′-d]pyrrol fused nonfullerene acceptors enabling over 13% efficiency for organic solar cells. *Advanced Materials* 30(16), 1707150. doi: 10.1002/adma.201707150.

Sun, Y., Li, G., Wang, L., Huai, Z., Fan, R., Huang, S., Fu, G. & Yang, S. (2018b) Simultaneous enhancement of short-circuit current density, open circuit voltage and fill factor in ternary organic solar cells based on PTB7-Th:IT-M:PC71BM. *Solar Energy Materials and Solar Cells* 182, 45–51. doi: 10.1016/j.solmat.2018.03.014.

Tang, Z., Tress, W. & Inganäs, O. (2014) Light trapping in thin film organic solar cells. *Materials Today* 17(8), 389–396. doi: 10.1016/j.mattod.2014.05.008.

Tvingstedt, K., Andersson, V., Zhang, F. & Inganäs, O. (2007) Folded reflective tandem polymer solar cell doubles efficiency. *Applied Physics Letters* 91, 123514. doi: 10.1063/1.2789393.

Veldman, D., Meskers, S.C.J. & Janssen, R.A.J. (2009) The energy of charge-transfer states in electron donor–acceptor blends: insight into the energy losses in organic solar cells. *Advanced Functional Materials* 19, 1939–1948. doi: 10.1002/adfm.200900090.

Voroshazi, E., Verreet, B., Aernouts, T. & Heremans, P. (2011) Long-term operational lifetime and degradation analysis of P3HT: PCBM photovoltaic cells. *Solar Energy Materials and Solar Cells* 95, 1303–1307. doi: 10.1016/j.solmat.2010.09.007.

Wadsworth, A., Bristow, H., Hamid, Z., Babics, M., Gasparini, N., Boyle, C.W., Zhang, W., Dong, Y., Thorley, K.J., Neophytou, M., Ashraf, R.S., Durrant, J.R., Baran, D. & McCulloch, I. (2019) End group tuning in acceptor–donor–acceptor nonfullerene small molecules for high fill factor organic solar cells. *Advanced Functional Materials* 29, 1808429. doi: 10.1002/adfm.201808429.

Wang, Y., Cong, H., Yu, B., Zhang, Z., & Zhan, X. (2017) Efficient inverted organic solar cells based on a fullerene derivative-modified transparent cathode. *Materials* 10(9), 1064. doi: 10.3390/ma10091064.

Wang, J., Xie, S., Zhang, D., Wang, R.., Zheng, Z., Zhou, H. & Zhang, Y. (2018) Ultra-narrow bandgap non-fullerene organic solar cells with low voltage losses and a large photocurrent. *Journal of Materials Chemistry A* 6, 19934–19940. doi: 10.1039/C8TA07954D.

Weng, K., Li, C., Bi, P., Sook Ryu, H., Guo, Y., Hao, X., Zhao, D., Li, W., Woo, H.Y. & Sun, Y. (2019) Ternary organic solar cells based on two compatible PDI-based acceptors with an enhanced power conversion efficiency. *Journal of Materials Chemistry A* 7(8), 3552–3557. doi: 10.1039/C8TA12034J.

Xiao, Z., Geng, X., He, D., Jia, X. & Ding, L. (2016) Development of isomer-free fullerene bisadducts for efficient polymer solar cells. *Energy & Environmental Science* 9(6), 2114–2121. doi: 10.1039/C6EE01026A.

Xiao, Z., Duan, T., Chen, H., Sun, K. & Lu, S. (2018) The role of hydrogen bonding in bulk-heterojunction (BHJ) solar cells: a review. *Solar Energy Materials and Solar Cells* 182, 1–13. doi: 10.1016/j.solmat.2018.03.013.

Xie, J., Wang, X., Wang, S., Ling, Z., Lian, H., Liu, N., Liao, Y., Yang, X., Qu, W., Peng, Y., Lan, W. & Wei, B. (2019) Solution-processed ZnO/MoS$_2$ quantum dots electron extraction layer for high performance inverted organic photovoltaics. *Organic Electronics* 75, 105381. doi: 10.1016/j.orgel.2019.105381.

Xiong, J., Jin, K., Jiang, Y., Qin, J., Wang, T., Liu, J., Liu, Q., Peng, H., Li, X., Sun, A., Meng, X., Zhang, L., Liu, L., Li, W., Fang, Z., Jia, X., Xiao, Z., Feng, Y., Zhang, X., Sun, K., Yang, S., Shi, S. & Ding, L. (2019) Thiolactone copolymer donor gifts organic solar cells a 16.72% efficiency. *Science Bulletin* 64(21), 1573–1576. doi: 10.1016/j.scib.2019.10.002.

Xu, C., Wright, M., Ping, D., Yi, H., Zhang, X., Mahmud, M.A., Sun, K., Upama, M.B., Haque, F. & Uddin, A. (2018) Ternary blend organic solar cells with a non-fullerene acceptor as a third component to synergistically improve the efficiency. *Organic Electronics* 62, 261–268. doi: 10.1016/j.orgel.2018.08.029.

Yao, H., Yu, R., Shin, T.J., Zhang, H., Zhang, S., Jang, B., Uddin, M.A., Woo, H.Y. & Hou, J. (2016) A wide bandgap polymer with strong π-π interaction for efficient fullerene-free polymer solar cells. *Advanced Energy Materials* 6(15), 1600742. doi: 10.1002/aenm.201600742.

Yu, G., Gao, J., Hummelen, J.C., Wudl, F. & Heeger, A.J. (1995) Polymer photovoltaic cells: enhanced efficiencies via a network of internal donor-acceptor heterojunctions. *Science* 270, 1789–1791. doi: 10.1126/science.270.5243.1789.

Yu, J., Xi, Y., Chueh, C.-C., Xu, J.-Q., Zhong, H., Lin, F., Jo, S.B., Pozzo, L.D., Tang, W. & Jen, A.K.-Y. (2017) Boosting performance of inverted organic solar cells by using a planar coronene based electron-transporting layer. *Nano Energy* 39, 454–460. doi: 10.1016/j.nanoen.2017.07.031.

Yu, Y.-Y., Tseng, C., Chien, W.-C. & Chen, C.-P. (2019a) Interface modification layers for high-performance inverted organic photovoltaics. *Organic Electronics* 69, 20–25. doi: 10.1016/j.orgel.2019.02.009.

Yu, R., Yao, H., Cui, Y., Hong, L., He, C. & Hou, J. (2019b) Improved charge transport and reduced nonradiative energy loss enable over 16% efficiency in ternary polymer solar cells. *Advanced Materials* 31(36), 1902302. doi: 10.1002/adma.201902302.

Yuan, J., Zhang, Y., Zhou, L., Zhang, G., Yip, H.-Y., Lau, T.-K., Lu, X., Zhu, C., Peng, H., Johnson, P.A., Leclerc, M., Cao, Y., Ulanski, J., Li, Y. & Zou, Y. (2019) Single-junction organic solar cell with over 15% efficiency using fused-ring acceptor with electron-deficient core. *Joule* 3(4), 1140–1151. doi: 10.1016/j.joule.2019.01.004.

Zhang, F.J., Zhao, D.W., Zhuo, Z.L., Wang, H., Xu, Z. & Wang, Y.S. (2010) Inverted small molecule organic solar cells with Ca modified ITO as cathode and MoO$_3$ modified Ag as anode. *Solar Energy Materials and Solar Cells* 94(12), 2416–2421. doi: 10.1016/j.solmat.2010.08.031.

Zhang, F., Xu, X., Tang, W., Zhang, J., Zhuo, Z., Wang, J., Wang, J., Xu, Z. & Wang, Y. (2011) Recent development of the inverted configuration organic solar cells. *Solar Energy Materials and Solar Cells* 95(7), 1785–1799. doi: 10.1016/j.solmat.2011.02.002.

Zhang, G., Zhang, K., Yin, Q., Jiang, X.-F., Wang, Z., Xin, J., Ma, W., Yan, H., Huang, F. & Cao, Y. (2017) High-performance ternary organic solar cell enabled by a thick active layer containing a liquid crystalline small molecule donor. *Journal of American Chemical Society* 139(6), 2387–2395. doi: 10.1021/jacs.6b11991.

Zhang, H., Yao, H., Hou, J., Zhu, J., Zhang, J., Li, W., Yu, R., Gao, B., Zhang, S. & Hou, J. (2018) Over 14% efficiency in organic solar cells enabled by chlorinated nonfullerene small-molecule acceptors. *Advanced Materials* 30(28), 1800613. doi: 10.1002/adma.201800613.

Zhang, C., Wang, W., Zhao, F., Pan, R., Zhang, J., Yu, H. & Li, J. (2020) Essential relation of spin states, trap states and photo-induced polarization for efficient charge dissociation in a polymer-nonfullerene based organic photovoltaic system. *Nano Energy* 78, 105324. doi: 10.1016/j.nanoen.2020.105324.

Zheng, Z., Awartani, O.M, Gautam, B., Liu, D., Qin, Y., Li, W., Bataller, A., Gundogdu, K., Ade, H. & Hou, J. (2017) Efficient charge transfer and fine-tuned energy level alignment in a THF-processed fullerene-free organic solar cell with 11.3% efficiency. *Advanced Materials* 29(5), 1604241. doi: 10.1002/adma.201604241.

Zhou, Z., Xu, S., Song, J., Jin, Y., Yue, Q., Qian, Y., Liu, F., Zhang, F. & Zhu, X. (2018) High-efficiency small-molecule ternary solar cells with a hierarchical morphology enabled by synergizing fullerene and non-fullerene acceptors. *Nature Energy* 3, 952–959. doi: 10.1038/s41560-018-0234-9.

Chapter 3

Dye-sensitized Solar Cells

3.1 INTRODUCTION

As photoelectrochemical devices, dye-sensitized solar cells (DSSCs) differ in terms of structure and operation principle from traditional silicon solar cells belonging to the first generation of photovoltaic technology. In silicon cells, the main part in which photovoltaic conversion occurs is the solid semiconductor p-n junction, while dye cells consist of two electrodes immersed in the electrolyte. In such photoelectrochemical devices, the fundamental stages of photovoltaic conversion of energy take place in different media: dye molecules are responsible for absorption of light, charge separation occurs at dye/semiconductor and dye/electrolyte interfaces, and transport of photoelectrons occurs in the layer of semiconductor nanoparticles.

The invention of DSSCs by Brian O'Regan and Michael Grätzel in 1991 was based on the long-term development of photoelectrochemistry, studies on solar energy conversion and achievements of nanotechnology. The first step was the observation of the photovoltaic effect made by Alexandre Edmond Becquerel in the photoelectrochemical system, at the interface between solid state and electrolyte, almost 200 years ago. The important contribution was also the research in the field of photochemistry, especially on the usage of the solar energy in artificial photosynthesis and promotion of the pioneering idea of fossil fuels replacement with renewable energy sources by Giacomo Camician. Further achievements, such as the introduction of sensitization method in black/white and then color photography, successfully performed photolysis of water, and the development of nanotechnology undoubtedly laid the foundation for the development of DSSCs.

In this chapter, the basics of DSSC operation and the particular parts of the dye photocells are described, including alternative materials that can substitute typical coverage of nanostructured electrodes and liquid electrolyte. The diversity of sensitizing dyes divided into metal-based complexes, metal-free compounds and dyes of natural origin extracted from plants is also presented.

3.2 STRUCTURE AND BASICS OF DSSC OPERATION

The idea of electric energy generation in a photoelectrochemical cell illuminated with visible light has quite a long history (Heller, 1981). At present, this type of device has been dominated by the design based on a semiconductor electrode photosensitized by a layer of light-absorbing molecules, known as the dye-sensitized solar cell (DSSC) or the Grätzel cell (O'Regan & Grätzel, 1991; Hagfeldt et al., 2010).

The conventional structure of a DSSC consists of the following components: (1) photoanode made of glass substrate covered by TCO (transparent conductive oxide) and the layer of titanium dioxide nanoparticles sensitized with the dye, (2) the counter electrode and (3) electrolyte inserted between the electrodes. The particular parts of the cell are indicated in Figure 3.1.

The substrate of photoanode is usually glass with deposited TCO layer, which should be transparent for the impinging light and characterized by good conductivity. Fluorine tin oxide (FTO) works best as a transparent conductive material in DSSC due to constant, low resistivity up to 600°C. Maintenance of good conductivity at high temperatures is crucial, since the sintering of TiO_2 deposited on the glass electrode requires annealing at 450°C. Therefore, FTO outperforms the popular indium tin oxide (ITO), the resistance of which increases with temperature. Another important advantage of FTO is lower price compared to ITO (Abrol et al., 2021).

One of the key components for energy conversion efficiency in DSSC is the n-type TiO_2 mesoporous film. Two well-established approaches can be applied to obtain a thin layer of titanium dioxide nanoparticles on the conductive glass substrate: sol-gel synthesis based on spreading of TiO_2 layer followed by calcination at 450°C–550°C, or the preparation of TiO_2 paste that is then deposited by the doctor blade, spin coating or screen printing methods. The common practice of TiO_2 paste application may be realized with the use of different commercially available TiO_2 nanopowders. They present a wide range of parameters, from the most frequently exploited P25 from Degussa® (now Evonik®) with a grain size of about 20–25 nm and containing at least 80% of anatase as well as 20% of rutile crystalline structure, to pure anatase TiO_2 nanopowder with a particle size of about 5 nm available from Aldrich®. The paste is prepared by grinding the TiO_2 powder in a mortar with acetic acid, water and detergent in order to obtain smooth consistency (Ito et al., 2007; Zama et al., 2017). Then, the TiO_2 paste has to be spread over the FTO substrate on glass, typically covering a rectangular area of $1 cm^2$. Further treatment includes drying in the air and annealing at 450°C–550°C for 30 minutes in order to sinter the nanoparticles and provide electric contact between them. After thermal treatment, the TiO_2 mesoporous layer becomes semitransparent and ready to be sensitized with a dye.

Titanium dioxide is a wide-bandgap semiconductor (E_g=3.2 eV) absorbing the solar radiation only below 380 nm. The ability of ultraviolet (UV) absorption by TiO_2 is useful in various applications; however, a wide range of light absorption is needed for effective photovoltaic energy conversion. In order to utilize the visible range of solar radiation, a sensibilization process of TiO_2 by the dyes has to be employed as the next step in the preparation of photoanode. The mechanism of dye sensitization and its role in photoelectrochemistry was established by Gerischer and Tributsch in experiments with ZnO (Gerischer et al., 1968). In a DSSC, the dye molecules are adsorbed on the nanoparticles of titanium dioxide, which form a mesoporous layer offering a large active surface. In practice, the adsorption takes place after the immersion of electrode in a dye solution, which penetrates the pores in TiO_2 layer, and results in bonding of the dye molecules to the nanoparticles. If it is needed, the amount of the adsorbed dye can be estimated by removing it from the electrode by splashing it with diluted NaOH and the measurement of absorption spectrum of the obtained solution (Liu et al., 2012). The typical features of a photoanode are presented in Table 3.1.

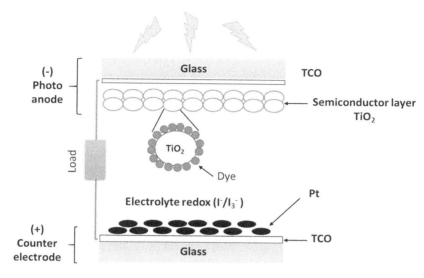

Figure 3.1 The schematic structure of a dye-sensitized solar cell. (Reprinted from Yahya et al. (2021). Copyright with permission from Elsevier.)

Another important element of the DSSC device is the electrolyte that mediates the inner charge carrier transport between the electrodes. The electrolyte should not react with the elements of the cell, and its absorption spectrum should not overlap with the absorption spectrum of the sensitizer.

The counter electrode in a DSSC is usually glass with conducting TCO film covered with a very thin layer of platinum black, i.e., Pt nanoparticles providing a highly developed surface. The Pt film is obtained by drying of a drop of chloroplatinic acid solution in ethanol on FTO layer and annealing of the platinized plates at 380°C for 30 min.

In order to stabilize the structure of the photoelectrochemical dye cell, a thermoplastic sealing film is placed between the photoanode and counter electrode, then pressed and heated to the appropriate temperature. The injection of the electrolyte into the ~0.1 mm gap between the electrodes is performed through a small hole in the counter electrode.

Table 3.1 Data of a typical photoanode

Parameter	Typical value
FTO sheet resistance	15 ohm/square
FTO transmittance	80% in visible light
TiO_2 layer thickness	5–15 µm
TiO_2 layer surface area	50–250 m^2/g
Porosity of TiO_2 layer	50%–65%
Pore size	15 nm
Size of TiO_2 nanoparticles	15–20 nm

3.2.1 Work cycle of DSSC

When the photoanode is illuminated, the incident photons can be absorbed by the molecules of the sensitizing dye. Upon absorption of energy, electrons are promoted from the highest occupied molecular orbital (HOMO) to the lowest unoccupied molecular orbital (LUMO) of the dye molecule and become excited. Two types of electronic excited state of the dye are considered in the description of the electronic coupling of the dye with titanium oxide. One is a weak coupling case when the LUMO orbital of the dye just populated does not involve the orbitals of the adsorption site (unsaturated Ti atom), and the other is the strong coupling case when the overlap and orbital energies are matched, resulting in a new excited state of the dye molecule that involves one or several atoms of the oxide. In the latter case, the lowest energy band in the absorption spectrum of the dye is remarkably shifted to the red and broadened, or a quite new absorption band is observed with a much lowered transition energy. As a rule, the redshifted absorption band corresponds to a charge-transfer character of the electronic transition, which was directly confirmed experimentally using the Stark effect spectroscopy (Krawczyk et al., 2018; Nawrocka et al., 2009) and by quantum chemical analyses of the excited state of the dye-oxide system (Sánchez-de-Armas et al., 2011). This is not always the case, and a strong coupling does not seem to be necessary for high overall effectiveness of the DSSC. Although the charge-transfer character of the excited state could be acquired immediately during the electronic transition, it remains unclear whether it could enhance the effectiveness of electron transfer into the oxide and what its effect could be on the overall electron transport within the photoanode. On the other hand, the kinetics of electron transfer at the molecule-solid interface in the weak coupling case has been described (Creutz et al., 2005, 2006) within the framework of the Marcus theory, which was originally formulated for molecular redox reactions.

After the charge separation at the oxide surface, the subsequent stages of charge transport take place in the direction of decreasing electron energy, as depicted in Figure 3.2. The electrons pass the layer of nanoparticles that are sintered and diffuse through the conductive coverage of the electrode toward the external electric contact. Meanwhile, the electrolyte takes part in redox reactions: the oxidized dye is regenerated by electrolyte, the electrolyte recovers the electron by reduction at the counter electrode, and the working cycle is closed. The operational sequence depicted in Figure 3.1 includes the following stages (Sharma et al., 2018):

- Under illumination, a photon is absorbed and the molecule of sensitizer (S) becomes excited:
  $S + h\nu \rightarrow S^*$.
- The electron is transferred from the LUMO level of the sensitizer into the conduction band of semiconductor leaving the oxidized dye molecule:
  $S^* \rightarrow S^+ + e^-$.
- The injected electrons are transported to the FTO layer, and then through the electric circuit outside the cell, the electrons travel to the counter electrode.
- The dye molecule regenerates with the electron drawn from the electrolyte:
  $S^+ + e^- \rightarrow S$.

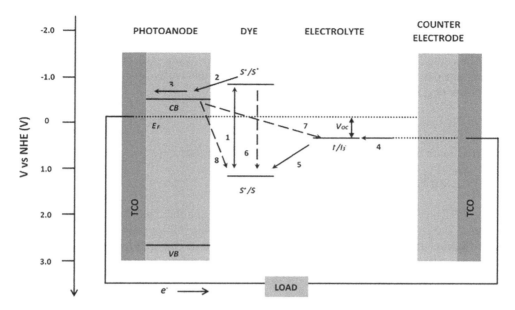

Figure 3.2 Energy level scheme of the dye-sensitized solar cell.

- The triiodide I_3^- ions take electrons from the counter electrode and become reduced to iodide:
$I_3^- + 2e^- \rightarrow 3I^-$.

The course of the presented processes depends on the proper configuration of energy levels. As shown in Figure 3.2 for the simpler case of weak coupling, the energy of the LUMO level of the dye molecule has to be higher than the bottom of semiconductor conduction band and the redox potential of the electrolyte has to lie above the HOMO level of the dye, but below the TiO_2 conduction band. The duration time of the subsequent processes and the kinetics of the whole working cycle are also important factors influencing the final performance of the cell. Commonly, the relaxation of the excited state in the sensitizer molecule takes 60 ns. The faster process taking 50 fs to 1.7 ps is the injection of the electron from the excited state to conduction band of TiO_2, followed by its drifting through TiO_2 or by the return from the conduction band to the dye HOMO level within nano- to milliseconds, which means the deleterious recombination (Karim et al., 2019). The regeneration of the photosensitizer by electrolyte proceeds very quickly, i.e., in 10 ns, which is important to maintain and complete the working cycle. The evaluation of DSSC performance is based on measurements of current-voltage (I-V) characteristic and incident photon-to-current conversion efficiency (IPCE) (exemplary characteristics are shown in Section 3.3.2, Figure 3.5). The typical parameters of the solar cell, such as short-circuit current (I_{SC}), open-circuit voltage (V_{OC}), maximum power and fill factor (FF) can be determined from the I-V characteristic. One of the most important parameters of the overall performance is the energetic efficiency, which is calculated according to the formula common for photovoltaic cells:

$$\eta = \frac{V_{OC} I_{SC} FF}{P_{in}} \times 100\%, \tag{3.1}$$

where P_{in} – the power of incident light.

The value of IPCE expressing the ratio of number of photoelectrons to the number of incident photons for a given wavelength λ is the spectrally resolved external quantum efficiency, described by the following formula:

$$IPCE(\lambda) = 1240 \times \frac{I_{SC}}{P_{in} \lambda}. \tag{3.2}$$

The IPCE can also be determined according to the relation:

$$IPCE = \phi_1 \phi_2 LHE,$$

where LHE – the light harvesting efficiency, Φ_1 – the quantum yield of charge injection, and Φ_2 – the charge collection efficiency. In DSSC, the photovoltage is determined by the difference between the Fermi level in the semiconductor conduction band and the redox potential of the electrolyte. The photocurrent value depends on LHE and efficiency of charge injection and collection.

Experimental measurements of I-V characteristics of a DSSC can be complemented by the models developed to gain scientific insight into the diversity of phenomena observed in dye cells, which enable the identification of the potentially crucial internal processes and the potentially promising materials. Different approaches are presented by the macroscopic diffusion model (Gagliardi et al., 2011), trap-limited diffusion model (Wang et al., 2014) as well as equivalent circuit single-diode model (Tian et al., 2011). The latter model, usually applied to silicon solar cells, is convenient since it does not consider numerous physical and chemical parameters. Instead, each single element of the electric circuit reflects the given physicochemical process.

3.2.2 Inverted configuration of DSSC

The substitution of n-type TiO_2 with p-type semiconductor as the sensitized coverage of photoelectrode in DSSCs is also possible. In this approach, the dye cell operates in the inverse mode; however, the basic working principle is similar. Upon excitation of the dye molecule, the hole is injected into the valence band of semiconductor, while the electron is transferred from the valence band into the half-empty orbital just left by the excited electron. The excited electron is transferred from the dye to the redox mediator and then to the counter electrode. The relative location of energy levels and the directions of charge transfer in p-type DSSC are depicted in Figure 3.3. Most often, a nanostructured nickel oxide (NiO) film is employed as p-type coverage of photoanode; however, the electrical conductivity of NiO is much lower than that of n-type semiconductors. The drawback of this solution is also the occurrence of cracks in the NiO layer, which deteriorate the optical quality and limit the absorption of light (Gong et al., 2017). The overall performance of the inverted DSSC is not satisfactory, since the basic parameters such as conversion efficiency, photovoltage and current densities are generally poorer than in n-type cells.

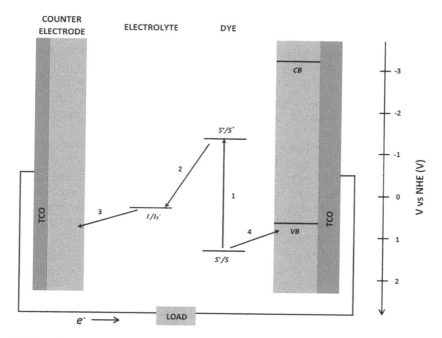

Figure 3.3 The scheme of energy levels in p-type DSSCs (DSSC of inverted structure). The steps of operation: 1 – photoexcitation of the dye molecule, 2 – transfer of electron to the oxidized species in the electrolyte, 3 – transfer of electron to the counter electrode and 4 – injection of a hole from the photoexcited dye to valence band of semiconductor.

3.3 REVIEW OF MATERIALS USED IN DSSC

3.3.1 Mesoporous semiconductor layer

Titanium dioxide is the most commonly used semiconductor material in dye cells. The important features of TiO_2, as an electron transport material applied in DSSCs, include satisfying electronic properties, resistance to photocorrosion, non-toxicity and low cost. In general, three crystalline forms of TiO_2 are distinguished: anatase, rutile and brookite. Among them, anatase is preferred for usage in DSSCs due to the optimum bandgap of 3.2 eV, the value of electron diffusion coefficient better than in rutile and easier preparation methods than that of brookite. Moreover, the thermodynamic properties of these polymorphs prefer the anatase crystalline structure when small-sized nanoparticles are formed.

The size of TiO_2 nanoparticles should be optimal to combine sufficient adsorption area for the dye and low number of grain boundaries working as traps of electrons. Various methods such as sol-gel, hydrothermal, microwave-assisted and template methods can provide the TiO_2 nanoparticles of 10–20 nm size, appropriate for applications in DSSCs (Karim et al., 2019). The nanoparticles of such size form a mesoporous structure of TiO_2 layer (Figure 3.4) with pores of similar size that are filled with the electrolyte.

Apart from the geometrical size of nanoparticles, the kind of crystalline facets prevailing on nanoparticle surface can be a factor of high relevance, taking into account the role of TiO_2 in dye cells. The significance of different facets was analyzed in the literature in various aspects relevant for DSSC functioning, e.g., with respect to the effectiveness of dye molecules adsorbed on different sites or peculiarities of electron transfer.

To achieve the exposition of desired crystalline facets, selective enhancement of particular crystallization routes, etching or capping ability of some chemical agents has been exploited. For instance, nanocrystals with anatase structure can be synthesized in the form of flat plates with the (001) facet exposed on the largest outer planes, using the hydrothermal method with the presence of hydrofluoric acid (Yang et al., 2008; Li et al., 2015). The product is a bipyramid (octahedron) with the tops strongly truncated along the (001) facets with square shape parallel and near to their common base. An opposite structure of nanocrystals with sharp bipyramidal shape exposing exclusively the oblique (101) facets is obtained when hydrothermal synthesis is performed in the presence of high amount of KOH (Li et al., 2015; Amano et al., 2009). Another approach for achieving the synthesis of preferable crystalline facets in TiO_2 nanocrystals is based on the solvothermal method using oleylamine, oleic acid or their mixtures in different proportions (Zhuang et al., 2015; Liu et al., 2009). Recently, anatase nanocrystals have been obtained from supercritical drying of a titanium hydroxide sol without using surface-specific coverage (Tobaldi et al., 2016). Besides, these nanocrystals were of 10 nm size, more than an order of magnitude smaller than those obtained using capping agents, thus revealing a larger specific surface in adsorption.

The performance of the faceted TiO_2 nanocrystals in DSSCs was found to be 43% better than that of the standard spherical P25 nanoparticles (Fang et al., 2011). Similarly, the DSSC sensitized with the MK-2 dye was the highest when (010) was the prevailing facet and yielded the overall cell efficiency of 6.1%, while 3.2% and 2.6%

Figure 3.4 Scanning electron microscope image of (a) TiO2 mesoporous layer and (b) TiO_2 layer with adsorbed dye molecules. (Reprinted from Kabir et al. (2019). Copyright with permission from Elsevier.)

were obtained, respectively, for (101) and (001) TiO_2 (Li et al., 2015). This sequence of efficiency was remarkably different from that in photocatalysis, where the (001) facet turned out to be several times more effective. These observations point to the high specificity of the facet-dependent effects, which may depend also on mean particle size and their state of aggregation, which was different in the experiments mentioned above. The source of these discrepancies may be sought in the details of the electronic processes that prevail at different surface planes. It was proposed (Yu et al., 2014) that the different facets in a nanocrystal display a coordinated sequence of events as a result of segregation of photogenerated electrons on the (101) facet and holes on the (001) facet. Recently, this hypothesis found support in a study of photogenerated changes in electron densities measured by XPS (Liu et al., 2016).

Another approach to improve the properties of TiO_2 and the performance of DSSCs, widely employed in research on photocatalysis, is doping of the TiO_2 nanocrystals by small amounts of nonmetallic or metallic dopants. Among them, nitrogen was the most frequently used, but also carbon, sulfur and phosphorus as well as many common metals such as Fe, Ni, Cu, Zn, Li, Ag, Sn, Sb, V, Cr, Co, Mo and rare earth elements were introduced (Kumar & Devi, 2011). The nitrogen is also introduced in TiO_2 applied in photocatalysis (Asahi et al., 2001) due to improvement in optical absorption and photocatalytic activities. The dye cell based on S-doped TiO_2 nanoparticles, in which sulfate groups generate active surface oxygen vacancies and act in the electrons transfer to the TiO_2 conduction band, demonstrated a PCE of 6.91% (Mahmoud et al., 2018). The metal dopants in DSSCs can change the electronic properties of the semiconductor, e.g., by increasing the number of charge carriers and introducing new electronic energy levels in semiconductor energy gap. The metal ions introduced in the anatase crystalline structure of TiO_2 change their valence state either permanently or by capturing the electrons photogenerated in the conduction band. Depending on the position relative to the surface, they can thus influence the reactivity of the surface. Various dopants can be added at the stage of chemical synthesis of TiO_2 nanocrystals, or by physical methods such as ion implantation, anodization and thermal oxidation (Yang et al., 2011).

The introduction of other materials into TiO_2 also includes preparation of TiO_2 composites with carbon nanomaterials such as graphene or single-walled and multi-walled carbon nanotubes, which is a very promising solution since carbon nanostructures improve electron transport and reduce recombination. Additionally, the position of graphene work function (−4.42 eV) is beneficial for the cascade electron transfer from TiO_2 to conductive coverage of the photoanode (Tang et al., 2010; Barberio et al., 2016; Karthikeyan et al., 2017; Low & Lai, 2018).

The crucial feature of TiO_2 layer employed in DSSCs is electrical conductivity, which can be improved by the employment of nanoparticles of new shapes. The developed and tested shapes of TiO_2 nanostructures in DSSCs include hollow spheres, nanotubes, nanowires and nanofibers synthetized by different methods. These new forms of TiO_2 nanostructures can provide efficiencies exceeding 10%, owing to good light scattering ability (e.g., in the case of hollow spheres) and improved light absorption in the visible range (e.g., in the case of nanofibers) (Karim et al., 2019).

Electronic and photocatalytic properties of TiO_2 can also be improved by hydrogenation, which reduces the bandgap and introduces additional mid-bandgap states. Another novel approach is the protonation of TiO_2 by acid treatment aimed at shifting the conduction band and better electronic coupling with the sensitizer (Gong et al., 2017). The modifications of TiO_2 layer also include the reduction of oxygen vacancies, which introduce surface states trapping the electrons. The limitation of deleterious oxygen vacancies can be realized by plasma treatment with oxygen, nitrogen or hydrogen on mesoporous TiO_2 (Gong et al., 2017).

In another promising solution, the enhancement of DSSC efficiency can be obtained with an increase in light path length in TiO_2 layer by the scattering of light penetrating the cell. To this aim, large TiO_2 nanoparticles can be incorporated in the cell structure. Experimental research demonstrated that mixing of large, scattering light nanoparticles of 400 nm size with the normally prepared TiO_2 paste (Chiba et al., 2006a) broadened the IPCE to NIR region and increased the efficiency to over 11%. The alternative method is the deposition of two additional layers (semitransparent and nontransparent) of TiO_2, which led to a PCE of 11.43% (Chang et al., 2013).

One of the requirements for mesoporous semiconductor layer utilized in DSSCs is good adherence to the substrate and durability. In order to enhance the adherence of TiO_2 layer to FTO substrate, prior to its deposition, FTO can be coated by a thin compact TiO_2 or ZnO film (Kouhestanian et al., 2020). Such a thin additional blocking layer covers FTO completely and reduces the interface between FTO and electrolyte, which is a place where the electron recombination occurs.

Aside from titanium dioxide, other wide-bandgap semiconductors exhibiting sufficient electronic properties and resistance to photocorrosion were employed as well, in the form of mesoporous layer in DSSCs. Zinc oxide with a bandgap $E_g = 3.37$ eV and electronic features similar to TiO_2 was used in the form of nanoparticles, providing a PCE of over 5% (Chen et al., 2018). The usage of SnO_2 and WO_3 led to a conversion efficiency of 2.5%–3.5%. However, the trials with other semiconductors have never given the results comparable with TiO_2 (Karim et al., 2019).

3.3.2 Sensitizers

The dye in DSSCs acts as the sensitizer, which is a crucial component, as it absorbs the light and takes part in the whole cycle of electrons transfer. To serve well in DSSCs, the dye should fulfill several requirements:

- The principal recommended characteristic of the dye is the absorption of light in a wide range from UV-Vis to NIR region.
- The proper energy of HOMO and LUMO of the dye is required. The HOMO level should be placed lower than the redox potential of the electrolyte, and the LUMO should be higher than the bottom edge of semiconductor conduction band.
- The molecule of the dye should possess an anchoring group that provides a durable bond with the surface of TiO_2 nanoparticles.
- The adsorption of the dye should provide thorough coverage of TiO_2 nanoparticles to avoid the contact of a semiconductor with redox mediators in the electrolyte.
- It is preferable to have groups in the dye molecule, which enable a rapid charge transfer of the electrons to the conduction band of the semiconductor and holes to the redox mediator.

- The aggregation of dye molecules should be avoided in order to limit the recombination between TiO$_2$ and redox mediator and promote the stable adsorption. The unfavorable aggregation can be partly prevented by the presence of co-adsorbates.
- The oxidized state of the dye should be durable and resistant to high temperature.

A lot of research is being done to identify and synthesize the dyes matching these essential characteristics of the efficient sensitizer. The dyes investigated for application in DSSCs can be divided into synthetic metal-based or metal-free complexes and natural dyes.

3.3.2.1 Metal-based complexes

In the pioneering research by O'Regan and Grätzel on dye-sensitized solar devices, a trimeric ruthenium complex (RuL$_2$(μ-(CN)Ru(CN)L'$_2$)$_2$, where L=2,2'-bipyridine-4,4'-dicarboxylic acid and L'=2,2-bipyridine) was employed to provide the efficiency of 7.1%–7.9% (O'Regan and Grätzel, 1991). Since then, numerous other ruthenium dyes were used in DSSCs; among them, the dye coded as N3 (cis-bis(isothiocyanato)-bis(4,40-dicarboxylic acid-2,20-bipyridine)Ru(II)) exhibited an IPCE of up to 800 nm, sufficient life time of the excited state and power conversion efficiency (PCE) of 10% in the early stage of research (Nazeeruddin et al., 1993). In further investigations, the N3 dye exhibited ultrafast electron injection following the absorption of photon and attained a PCE of 11.18%, which is record efficiency for ruthenium sensitizers (Nazeeruddin et al., 2005). The PCE of 11.1%, close to that record, was also achieved by another ruthenium dye – black dye, designated as N749 (tris(isothiocyanato)-2,20,2"-terpyridyl-4,40,4"-tricarbolylate)Ru(II) complex) (Chiba et al., 2006b). The ruthenium dyes present a broad range of light absorption and are most widely investigated as sensitizers in DSSCs; however, their absorption in near-infrared (NIR) region is weak, while high price and scarcity of ruthenium are additional disadvantages.

Other transition metals, such as osmium, rhenium and platinum, constitute an alternative to ruthenium as a metal component in the dye structure. Osmium polypyridines appeared as promising photosensitizers offering extended light absorption, which resulted in the efficiency over 8% (Wu et al., 2012); however, the main drawback was slow regeneration of the dye by electrolyte. The employment of other metals led to IPCE and PCE much lower than for ruthenium dyes (Yahya et al., 2021).

Porphyrins composed of four pyrrole units with a metal ion in central position are also attractive dyes for application in DSSCs. The inspiration for the application of porphyrins in DSSCs was their significant role in nature, especially in photosynthesis process. Porphyrins constitute a promising electron donor in DSSCs, since they offer favorable spectral characteristic, high stability, low cost and tunable absorption. The DSSC with zinc porphyrin dye coded as YD2-o-C8, co-sensitized with another organic dye Y123, reached a PCE of over 12% with cobalt redox shuttle electrolyte (Yella et al., 2011). The modification of porphyrin sensitizer, which leads to the more complex D-A-π-A structure, is an efficient way to tune the absorption spectrum (Jie et al., 2020). Such a structure is characteristic of zinc porphyrin

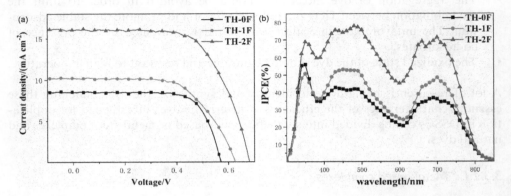

Figure 3.5 Current density vs. voltage characteristic (a) and IPCE plot (b) of dye cells sensitized with TH-0F, TH-1F and TH-2F. (Reprinted from Jie et al. (2020). Copyright with permission from Elsevier.)

sensitizers TH-0F (with no fluorine atom), TH-1F (with one fluorine atom) and TH-2F (with two fluorine atoms) with a benzothiadiazole unit, which demonstrated a redshift of absorption spectra due to the replacement of benzene by thiophene as a π bridge. The enhancement of light harvesting and charge transfer achieved in photocells sensitized with TH-2F led to a PCE of 6.98%. The current density vs. voltage characteristics and IPCE of dye cells sensitized with TH-0F, TH-1F and TH-2F are depicted in Figure 3.5.

3.3.2.2 Metal-free dyes

The metal-free organic compounds exhibiting a high extinction coefficient, low cost and long-term stability are an attractive alternative to dyes containing a metal ion. Organic dyes present various forms; among them, the very popular and simply synthesized structure is the donor-acceptor linkage through the π conjugated bridge (A-π-D). This type of structure can be designed to properly adjust the HOMO and LUMO levels and cover the wide light absorption range. The donor function can be performed by moieties such as indoline, triphenylamine, carbazole, fluorine and coumarin, while as an acceptor, ensuring the anchoring of sensitizer molecule to TiO_2, units such carboxylic acid, cyanoacrylic acid and rhodamine can be employed (Zhong et al., 2015; Arunkumar & Anbarasan, 2019). The examples of such sensitizers are coumarin-bearing triarylamine dyes (ZCJ-1, ZCJ-2 and ZCJ-3) exhibiting a high molar extinction coefficient and broad absorption range as visible in Figure 3.6. The new sensitizers combined excellent electronic properties and photoelectrochemical behavior of triarylamines with good absorption of visible light and long-term stability of coumarin derivatives. The usage of ZCJ-3 dye characterized by the highest extinction coefficient led to an efficiency of 6.24%. Typically, the carboxylic group is responsible for the formation of the TiO_2-dye molecule bond, which can be unidentate, chelate or bridging bidentate.

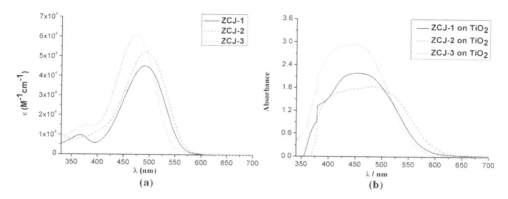

Figure 3.6 Molar extinction coefficient of triarylamine dyes (a) and absorption spectra of triarylamine dyes adsorbed on TiO$_2$ (b). (Reprinted from Zhong et al. (2015). Copyright with permission from Elsevier.)

3.3.2.3 *Natural dyes*

It is beneficial for the environment to replace synthetic dyes in DSSCs by natural ones due to their abundance in nature, non-toxicity and simple extraction methods. The group of natural dyes includes diverse structures such as anthocyanins, betalains, flavonoids, carotenoids, tannins and chlorophylls. Taking into account the accessibility, easy preparation without purification and large absorption coefficient, natural dyes are a valuable alternative that can be successfully used as a light harvester (Richhariya et al., 2017; Al-Alwani et al., 2018, 2020; Ferreira et al., 2020). However, the efficiencies of the dye cells sensitized with natural dyes are not satisfactory for some features of the natural dyes; for example, light degradation of curcumin or poor adsorption of phytyl and alkyl groups in the case of chlorophyll significantly deteriorates the performance of the cells.

The best performing natural dyes are anthocyanins (PCE of 2%), which offer the carbonyl and hydroxyl groups binding to the TiO$_2$ nanoparticles (Omar et al., 2020). Other groups of dyes, such as flavonoids, which are the compounds commonly occurring in leafs, stems, seeds, flowers or roots of the plants, were also investigated as sensitizers. Despite the existence of durable bonds with the surface of TiO$_2$ and fast electron transfer (Zdyb & Krawczyk, 2014; Zdyb & Krawczyk, 2016), the efficiencies of the dye cells with flavonoid sensitizers are below 2% (Omar et al., 2020).

3.3.3 Counter electrode

Typical coverage of counter electrode in DSSCs is platinum black comprised of nano-sized grains. Platinum is used due to its excellent photocatalytic properties, high conductivity and stability. Platinum coating on the counter electrode can be fabricated, e.g., by sputtering, electrochemical deposition or thermal decomposition. The goal is to obtain the layer of good adhesion and large surface area that can be achieved by deposition of platinum nanostructures in various forms such as nanoparticles, nanotubes, nanoflowers and nanowires (Wu et al., 2017). Besides high charge conductivity, a thin layer of the nanosized platinum also shows other desired properties: good transparency and resistance to corrosion in contact with ionic electrolyte. However, platinum is expensive and this disadvantage drives the search for various substitutes that can be applied as a coverage of counter electrode in DSSCs.

Carbon in the form of nanoparticles, nanotubes or graphene sheets, known for high surface area and very good electric conductivity, is a popular alternative material for counter electrode. Additionally, nanotubes and graphene have extraordinary mechanical and thermal properties as well as transparency of over 80%. The high transparency of the counter electrode, better than the typical 70% obtained for platinum coverage, is required in bifacial DSSC devices in which both front and rear light is converted. Graphene and graphene oxide present an outstanding catalytic activity; however, in the case of carbon nanotubes, this feature depends on their dimensions and structure. In an experimental research, multi-walled carbon nanotubes, grown in situ on metal nickel substrate, showed an excellent electrocatalytic performance and led to the 7.43% PCE of dye cell, higher than that for a typical counter electrode consisting Pt on FTO (Zheng et al., 2018).

Other materials that can be used in counter electrodes are conductive polymers, e.g., polyaniline (PANI) and polyethylenedioxythiophene (PEDOT). Besides the redox properties, the polymers exhibit lightweight, low cost, transparency and flexibility, which are important especially in the cells fabricated on elastic substrates. Polymers proved to be useful not only with iodide/iodine electrolyte, but also in copper- and cobalt-based electrolytes employed in highly efficient dye cells (Marchini et al., 2021).

In alternative counter electrodes, platinum can also be replaced by transition metals, e.g., tungsten, tantalum and niobium, which exhibit the capability of supplying or withdrawing electrons that is fundamental for their extraordinary catalytic activity. The application of transition metal compounds as a counter electrode, including sulfides, oxides, nitrides, carbides and selenides, allows achieving a PCE of around 5%–7% (Wu & Ma, 2018).

3.3.4 Electrolytes

The comprehensive reviews of the electrolytes employed in DSSCs are rather rare in the literature, although the role of the electrolyte in the cell is essential, since it is responsible for the transport of charge between the counter electrode and the photoelectrode as well as the regeneration of the oxidized dye molecule. The reliable electrolyte has to fulfill many requirements to serve in DSSCs. A good electrolyte should be non-reactive with other components of the cell and characterized by thermal and chemical stability and low absorption, which does not overlap with the absorption spectrum of the photoanode, as well as good contact at the interface with the porous surface of the photoelectrode (Mehmood et al., 2017). The electrolytes applied in dye-sensitized cells can be divided into three groups: liquid electrolytes, quasi-solid electrolytes and solid-state conductors.

The dye cells presenting the highest efficiencies were fabricated using liquid electrolytes; however, the stability is not the strength of these cells because of the possibility of electrolyte leakage. The most commonly applied liquid electrolyte containing iodide/triiodide redox couple was used in the conventional dye cells with ruthenium sensitizer N3 dye, which provided a PCE of 11.18% (Nazeeruddin et al., 2005). The traditional electrolyte with iodide/triiodide I^-/I_3^- redox couple in organic solvent remains the most popular choice, since it is known for good penetration of pores in photoelectrode and high conductivity. However, the iodide/triiodide electrolyte is corrosive and its redox level is high compared to the TiO_2 conduction band, thus lowering the open-circuit voltage V_{OC}.

Valuable alternatives to I^-/I_3^- redox mediator are cobalt, copper, ferrocene and nickel complexes, which are non-corrosive, are nonvolatile and exhibit low absorption of light (Iftikhar et al., 2019). These kinds of electrolytes also present a lower redox level than iodide/triiodide and, as a result, better match the energy of HOMO and LUMO of the dyes, which ensures higher values of the open-circuit voltage (Pradhan et al., 2018). The investigations show that the optimal adjustment of the redox potential and energy levels of dye molecules can significantly boost the efficiency of the cell. Cobalt redox shuttle is used as an electrolyte the most often; however, its high viscosity leads to the problems with mass transport and recombination losses. Due to the recent contribution of cobalt tris-bipyridine liquid electrolyte in record efficiency achieved for traditional device configuration, this kind of electrolyte is considered the most suitable one for application in dye cells (Kakiage et al., 2015). The usage of cobalt-based liquid electrolyte accompanied by co-sensitization with alkoxysilyl-anchor dye ADEKA-1 and a carboxy-anchor organic dye LEG4 (Kakiage et al., 2015) enabled attaining the record efficiency of 14.3%. The potential of cobalt electrolyte was also indicated in earlier studies on the cells sensitized with zinc porphyrin dye YD2-o-C8, which demonstrated an open-circuit voltage close to 1 V and efficiency of 12.3% (Yella et al., 2011).

The application of copper Cu(I/II) redox shuttle as the replacement of I^-/I_3^- is also a very promising solution, as it offers an excellent electron transfer rate, better mass transport compared to Co and fast dye regeneration, which results in an efficient DSSC performance. Through the application of a copper-based electrolyte, the PCE in the range of 11%–13.1% (Pradhan et al., 2018; Cao et al., 2018) and an open-circuit voltage over 1 V were obtained with the combination of two sensitizers (Freitag et al., 2017).

The ferrocenium/ferrocene redox couple (Fc/Fc^+), which is abundantly available similarly to Cu, has very beneficial properties, such as efficient regeneration of dye and low position of redox level (Fc^+/Fc has a redox potential of +0.403 V, Cu^{2+}/Cu^+ has +0.159 V, and Fe^{3+}/F^{2+} has +0.771 V), independently of the type of the solvent. However, in the cells with the electrolyte based on ferrocene compounds, rapid recombination between TiO_2 and dye occurs, so the achieved efficiencies are relatively low, i.e., around 5% (Daeneke et al., 2011).

Liquid electrolytes are usually based on two types of solvents. Organic solvents, e.g., acetonitrile (ACN) and valeronitrile, and other nitrile solvents belong to the first type of solvents that are of low toxicity (Iftikhar et al., 2019); however, they exhibit problems with leakage, and they are not suitable for flexible cells with conductive polymers as counter electrodes. The application of another type of solvent, e.g., methoxypropionitrile (MPN), resulted in the improvement of stability of dye cells. The cells based on ruthenium dye, platinum counter electrode and iodide/triiodide in MPN demonstrated long-term stability under exposure of light and elevated temperature of 80°C (Harikisun & Desilvestro, 2011).

Besides the electrolytes based on organic solvents, ionic liquids can be applied as the medium for the redox system in DSSCs, enabling the value of PCE of 6.9% (Xi et al., 2008). Ionic liquids (e.g., ammonium salts) present thermal and chemical stability, sufficient conductivity and low vapor pressure (Shi et al., 2008).

Liquid electrolytes offer numerous advantages that contribute to the achievement of the best efficiencies of energy conversion in DSSCs; however, except for ionic liquids, they are usually volatile and can leak out of the cell. These main drawbacks have a negative impact on the long-term stability of dye cells and are the reasons of problems with maintaining the initial efficiency. The best performing liquid iodide/triiodide electrolyte, which proved highly efficient performance, also creates additional issues, such as corrosion of Pt and dye desorption. Therefore, diverse kinds of gel electrolytes and solid-state electrolytes were developed to overcome the problems associated with liquids.

Gel electrolytes, in which the transport of anions and cations takes place owing to diffusion, gained interest due to excellent ionic conductivity and easy preparation methods. The improvement of mechanical and electrical properties obtained by the incorporation of some additives in gel electrolyte (metal oxide nanoparticles and pyridine derivatives) can enhance the performance of DSSC (Saidi et al., 2021); however, the main issue in the application of the hybrid gel material is its difficulty with the penetration into TiO_2 pores.

Much attention was also devoted to the development of all-solid-state DSSCs based on transporting holes solid p-type conductor in the role of the electrolyte, but the conversion efficiency of solid-state dye cells still remains below the results obtained with liquid electrolytes. The search for the solid substitutes of the liquid electrolyte includes the materials that can be classified to the following types: ionic conductors, inorganic hole-transport materials and organic hole-transport materials (Iftikhar et al., 2019).

The employment of liquid crystals, belonging to the first category, in which the transport of ionic reagents proceeds along the self-assembled lamellar layers of electrolyte, enabled the fabrication of efficient and stable DSSC, showing a PCE of 5.8% and extraordinary long-term stability for over 1000 h (Högberg et al., 2016).

Inorganic hole transport materials (e.g., NiO, CuI, CuSCN, CuO and Cu_2O) have very promising characteristics owing to low cost, simple preparation, chemical stability as well as high charge mobility (Li et al., 2012). The usage of copper iodide (CuI) and copper thiocyanate ($CsSnI_3$) led to the efficiencies of DSSCs of up to 7.4% (Sakamoto et al., 2012) and 8.5% (Chung et al., 2012), respectively. The excellent suitability of $CsSnI_3$, TiO_2 and N719 dye energy levels was a contributing factor.

Organic hole transport materials exhibit many desirable attributes, such as suitable hole mobility, easy preparation and deposition methods and cost-effectiveness. The employment of well-known p-type organic semiconductors in DSSCs attained quite good efficiencies of up to 5% for poly(3-hexylthiophene) (P3HT) (Chang et al., 2010) and 6.8% for PEDOT, which presents high hole conductivity and extraordinary stability (Kim et al., 2011). The further improvement of solid DSSC performance was achieved with spiro-MeOTAD doped with Co complex, resulting in an efficiency of 7.2% (Burschka et al., 2011) and 8% in cells sensitized with new quinoxaline-based D–A–π–A organic sensitizers (Li et al., 2017).

3.4 SUMMARY

The intensive investigations, carried out since the first report on the development of novel, promising dye-sensitized solar cells, led to their current energy conversion efficiency exceeding 14% and improvement of the long-term stability. Used from the very beginning, titanium dioxide proved to be the best electron transport material compared to other studied wide-bandgap semiconductors. Numerous works have demonstrated the advantage of using liquid cobalt-based electrolyte due to the beneficial position of the redox level in relation to the HOMO and LUMO of the sensitizers. However, the development of solid-state electrolytes providing better stability of the photocells seems to be critical for the prospective commercialization of dye cells.

REFERENCES

Abrol, S.A., Bhargava, C. & Sharma, P.S. (2021) Material and its selection attributes for improved DSSC. *Materials Today: Proceedings* 42, 1477–1484. doi: 10.1016/j.matpr.2021.01.312.

Al-Alwani, M.A.M., Ludin, N.A., Mohamad, A., Kadhum, A.H. & Mukhlus, A. (2018) Application of dyes extracted from Alternanthera dentata leaves and Musa acuminate bracts as natural sensitizers for dye-sensitized solar cells. *Spectrochimica Acta Part A* 192, 487–498. doi: 10.1016/j.saa.2017.11.018.

Al-Alwani, M.H.M., Hasan, H.A., Al-Shorgani, N.K.N. & Al-Mashaan, A.B.S.A. (2020) Natural dye extracted from Areca catechu fruits as a new sensitiser for dye-sensitized solar cell fabrication: optimisation using D-optimal design. *Materials Chemistry and Physics* 240, 122204. doi: 10.1016/j.matchemphys.2019.122204.

Amano, F., Yasumoto, T., Prieto-Mahaney, O.-O., Uchida, S., Shibayamad, T. & Ohtaniab, B. (2009) Photocatalytic activity of octahedral single-crystalline mesoparticles of anatase titanium(IV) oxide. *Chemical Communication* 17, 2311–2313. doi: 10.1039/B822634B.

Arunkumar, A. & Anbarasan, P.M. (2019) Optoelectronic properties of a simple metal-free organic sensitizer with different spacer groups: quantum chemical assessments. *Journal of Electronic Materials* 48, 1522–1530. doi: 10.1007/s11664-018-06912-x.

Asahi, R., Morikawa, T., Ohwaki, T., Aoki, K. & Taga, Y. (2001) Visible-light photocatalysis in nitrogen-doped titanium oxides. *Science* 293, 269–271. doi: 10.1126/science.1061051.

Barberio, M., Grosso, D.R., Imbrogno, A. & Xu, F. (2016) Preparation and photovoltaic properties of layered TiO_2/carbon nanotube/TiO_2 photoanodes for dye-sensitized solar cells. *Superlattices and Microstructures* 91, 158–164. doi: 10.1016/j.spmi.2016.01.012.

Burschka, J., Dualeh, A., Kessler, F., Baranoff, E., Cevey-Ha, N.-L., Yi, C., Nazeeruddin, M.K. & Gratzel, M. (2011) Tris(2-(1H-pyrazol-1-yl)pyridine)cobalt(III) as p-type dopant for organic semiconductors and its application in highly efficient solid-state dye-sensitized solar cells. *Journal of the American Chemical Society* 133(45), 18042–18045. doi: 10.1021/ja207367t.

Cao, Y., Liu, Y., Zakeeruddin, S.M., Hagfeldt, A. & Grätzel, M. (2018), direct contact of selective charge extraction layers enables high-efficiency molecular photovoltaics. *Joule* 2(6), 1108–1117. doi: 10.1016/j.joule.2018.03.017.

Chang, J.A., Rhee, J.H., Im, S.H., Lee, Y.H., Kim, H., Seok, S., Nazeeruddin, M.K., & Gratzel, M. (2010) High-performance nanostructured inorganic-organic heterojunction solar cells. *Nano Letters* 10, 2609–2612. doi: 10.1021/nl101322h.

Chang, Y.-J., Kong, E.-H., Park, Y.-C. & Jang, H.M. (2013) Broadband light confinement using a hierarchically structured TiO_2 multi-layer for dye-sensitized solar cells. *Journal of Materials Chemistry A* 1(34), 9707–9713. doi: 10.1039/C3TA11527E.

Chen, Y.-C., Li, Y.-J. & Hsu, Y.-K. (2018) Enhanced performance of ZnO-based dye-sensitized solar cells by glucose treatment. *Journal of Alloys and Compounds* 748, 382–389. doi: 10.1016/j.jallcom.2018.03.189.

Chiba, Y., Islam, A., Komiya, R., Koide, N. & Han, L.Y. (2006a) Conversion efficiency of 10.8% by a dye-sensitized solar cell using a TiO_2 electrode with high haze. *Applied Physics Leters* 88, 223505. doi: 10.1063/1.2208920.

Chiba, Y., Islam, A., Watanabe, Y., Komiya, R., Koide, N. & Han, L.Y. (2006b) Dye-sensitized solar cells with conversion efficiecy of 11.1%. *Japanese Journal of Applied Physics* 45(7L), 638–640. DOI: 10.1143/JJAP.45.L638.

Chung, I., Lee, B., He, J.Q., Chang, R.P.H. & Kanatzidis, M.G. (2012) All-solid-state dye-sensitized solar cells with high efficiency. *Nature* 485, 486–489. https://www.nature.com/articles/nature11067.

Creutz, C., Brunschwig, B.S. & Sutin, N. (2005) interfacial charge-transfer absorption: semi-classical treatment. *The Journal of Physical Chemistry B* 109(20), 10251–10260. doi: 10.1021/jp050259+.

Creutz, C., Brunschwig, B.S. & Sutin, N. (2006) Interfacial charge-transfer absorption: 3. Application to semiconductor-molecule assemblies. *The Journal of Physical Chemistry B* 110, 25181–25190. doi: 10.1021/jp063953d.

Daeneke, T., Kwon, T.-H., Holmes, A.B., Duffy, N.W., Bach, U. & Spiccia, L. (2011) High-efficiency dye-sensitized solar cells with ferrocene-based electrolytes. *Nature Chemistry* 3, 211–215. doi: 10.1039/C2EE21257A.

Fang, W.Q., Gong, X.-Q. &Yang, H.G. (2011) On the unusual properties of anatase TiO_2 exposed by highly reactive facets. *Journal of Physical Chemistry Letters* 2(7), 725–734. https:doi.org/10.1021/jz200117r.

Ferreira, F.C., Suresh Babua, R., de Barros, A.L.F., Raja, S., da Conceição, L.R.B. & Mattoso, L.H.C. (2020) Photoelectric performance evaluation of DSSCs using the dye extracted from different color petals of Leucanthemum vulgare flowers as novel sensitizers. *Spectrochimica Acta Part A: Molecular and Biomolecular Spectroscopy* 233, 118198. doi: 10.1016/j.saa.2020.118198.

Freitag, M., Teuscher, J., Saygili, Y., Zhang, X., Giordano, F., Liska, P., Hua, J., Zakeeruddin, S.M., Moser, J.-E. & Grätzel, M. (2017) Dye-sensitized solar cells for efficient power generation under ambient lighting. *Nature Photonics* 11, 372–378. https://www.nature.com/articles/nphoton.2017.60.

Gagliardi, A., Auf der Maur, M., Gentilini, D. & Carlo A. (2011) Simulation of dye solar cells: through and beyond one dimension. *Journal of Computational Electronics* 10, 424–436. doi: 10.1007/s10825-011-0377-4.

Gerischer, H., Michel-Beyerle, M.E., Rebentrost, F. & Tributsch, H. (1968) Sensitization of charge injection into semiconductors with large band gap. *Electrochimica Acta* 13(6), 1509–1515. doi: 10.1016/0013-4686(68)80076-3.

Gong, J., Sumathy, K., Qiao, Q. & Zhou, Z. (2017) Review on dye-sensitized solar cells (DSSCs): advanced techniques and research trends. *Renewable and Sustainable Energy Reviews* 68, 234–246. doi: 10.1016/j.rser.2016.09.097.

Hagfeldt, A., Boschloo, G., Sun, L., Kloo, L. & Pettersson, H. (2010) Dye-sensitized solar cells. *Chemical Reviews* 110, 6595–6663. doi: 10.1021/cr900356p.

Harikisun, R. & Desilvestro, H. (2011) Long-term stability of dye solar cells. *Solar Energy* 85(6), 1179–1188. doi: 10.1016/j.solener.2010.10.016.

Heller, A. (1981) Conversion of sunlight into electrical power and photoassisted electrolysis of water in photoelectrochemical cells. *Accounts of Chemical Research* 14(5), 154–162. doi: 10.1021/ar00065a004.

Högberg, D., Soberats, B., Yatagai, R., Uchida, S., Yoshio, M., Kloo, L., Segawa, H. & Kato, T. (2016) Liquid-crystalline dye-sensitized solar cells: design of two-dimensional molecular assemblies for efficient ion transport and thermal stability. *Chemistry of Materials* 28(18), 6493–6500. doi: 10.1021/acs.chemmater.6b01590.

Iftikhar, H., Sonai, G.G., Hashmi, S.G., Nogueira, A.F. & Lund, P.D. (2019) Progress on electrolytes development in dye-sensitized solar cells. *Materials* 12, 1998. doi: 10.3390/ma12121998.

Ito, S., Chen, P., Comte, P., Nazeeruddin, M.K., Liska, P., Péchy, P. & Grätzel, M. (2007) Fabrication of screen-printing pastes from TiO_2 powders for dye-sensitised solar cells. *Progress in Photovoltaics: Research and Applications* 15(7), 603–12. doi: 10.1002/pip.768.

Jie, J., Xu, Q., Yang, G., Feng, Y. & Zhang, B. (2020) Porphyrin sensitizers involving a fluorine-substituted benzothiadiazole as auxiliary acceptor and thiophene as π bridge for use in dye-sensitized solar cells (DSSCs). *Dyes and Pigments* 174, 107984. doi: 10.1016/j.dyepig.2019.107984.

Kabir, F., Sakib, S.N. & Matin, N. (2019) Stability study of natural green dye based DSSC. *Optik* 181, 458–464. doi: 10.1016/j.ijleo.2018.12.077.

Kakiage, K., Aoyama, Y., Yano, T., Oya, K., Fujisaw, J. & Hanaya, M. (2015) Highly-efficient dye-sensitized solar cells with collaborative sensitization by silyl-anchor and carboxy-anchor dyes. *Chemical Communications* 51(88), 15894–15897. doi: 10.1039/C5CC06759F.

Karim, N.A., Mehmood, U., Zahid, H.F. & Asif, T. (2019) Nanostructured photoanode and counter electrode materials for efficient Dye-Sensitized Solar Cells (DSSCs). *Solar Energy* 185, 165–188. doi: 10.1016/j.solener.2019.04.057.

Karthikeyan, V., Maniarasu, S., Manjunath, V., Ramasamy, E. & Veerappan, G. (2017) Hydrothermally tailored anatase TiO_2 nanoplates with exposed {111} facets for highly efficient dye-sensitized solar cells. *Solar Energy* 147, 202–208. doi: 10.1016/j.solener.2017.03.049.

Kim, J., Koh, J.K., Kim, B., Ahn, S.H., Ahn, N, Ryu, D.Y., Kim, J.H. & Kim, E. (2011) Enhanced performance of I2-free solid-state dye-sensitized solar cells with conductive polymer up to 6.8%. *Advanced Functional Materials* 21(24), 4633–4639. doi: 10.1002/adfm.201101520.

Kouhestanian, E., Mozaffari, S.A., Ranjbar, M. & Amoli, H.S. (2020) Enhancing the electron transfer process of TiO_2-based DSSC using DC magnetron sputtered ZnO as an efficient alternative for blocking layer. *Organic Electronics* 86, 105915. doi: 10.1016/j.orgel.2020.105915.

Krawczyk, S., Nawrocka, A. & Zdyb, A. (2018) Charge-transfer excited state in pyrene-1-carboxylic acids adsorbed on titanium dioxide nanoparticles. *Spectrochimica Acta Part A: Molecular and Biomolecular Spectroscopy* 198, 19–26. doi:10.1016/j.saa.2018.02.061.

Kumar, G. & Gomathi Devi, L. (2011) Review on modified TiO_2 photocatalysis under UV/visible light: selected results and related mechanisms on interfacial charge carrier transfer dynamics. *Journal of Physical Chemistry A* 115(46), 13211–13241. doi: 10.1021/jp204364a.

Li, M.-H., Yum, J.-H., Moon, S.-J. & Chen, P. (2012) Inorganic p-type semiconductors: their applications and progress in dye-sensitized solar cells and Perovskite solar cells. *Energies* 9, 331. doi: 10.3390/en9050331.

Li, C., Koenigsmann, C., Ding, W., Rudshteyn, B., Yang, K.R., Regan, K.P., Konezny, S.J., Batista, V.S., Brudvig, G.W., Schmuttenmaer, C.A. & Kim, J.-H. (2015) Facet-dependent photoelectrochemical performance of TiO_2 nanostructures: an experimental and computational study. *Journal of American Chemical Society* 137(4), 1520–1529. doi: 10.1021/ja5111078.

Li, X., Xu, B., Liu, P., Hu, Y., Kloo, L., Hua, J., Sun, L. & Tian, H. (2017) Molecular engineering of D–A–π–A sensitizers for highly efficient solid-state dye-sensitized solar cells. *Journal of Materials Chemistry A* 5(7), 3157–3166. doi: 10.1039/C6TA10673K.

Liu, B., Wang, X., Cai, G., Wen, L., Song, Y. & Zhao, X. (2009) Low temperature fabrication of V-doped TiO_2 nanoparticles, structure and photocatalytic studies. *Journal of Hazardous Materials* 169(1–3), 1112–1118. doi: 10.1016/j.jhazmat.2009.04.068.

Liu, K.-Y., Hsu, C.-L., Ni, J.-S., Ho, K.-C. & Lin, K.-F. (2012) Photovoltaic properties of dye-sensitized solar cells associated with amphiphilic structure of ruthenium complex dyes. *Journal of Colloid and Interface Science* 372(1), 73–9. doi: 10.1016/j.jcis.2012.01.004.

Liu, X., Dong, G., Li, S., Lu, G. & Bi, Y. (2016) Direct observation of charge separation on anatase TiO_2 crystals with selectively etched {001} facets. *Journal of American Chemical Society* 138, 2917–2920. doi: 10.1021/jacs.5b12521.

Low, F.W. & Lai, C.W. (2018) Recent developments of graphene-TiO_2 composite nanomaterials as efficient photoelectrodes in dye-sensitized solar cells: a review. *Renewable and Sustainable Energy Reviews* 82, 103–125. doi: 10.1016/J.RSER.2017.09.024.

Mahmoud, M.S., Akhtar, M.S., Mohamed, I.M.A., Hamdan, R., Dakka, Y.A. & Barakat, N.A.M. (2018) Demonstrated photons to electron activity of S-doped TiO_2 nanofibers as photoanode in the DSSC. *Material Letters* 225, 77–81. doi: 10.1016/J.MATLET.2018.04.108.

Marchini, E., Caramori, S., Bignozzi, C.A. & Carli, S. (2021) On the use of PEDOT as a catalytic counter electrode material in dye-sensitized solar cells. *Applied Sciences* 11, 3795. doi: 10.3390/app11093795.

Mehmood, U., Al-Ahmed, A., Al-Sulaiman, F.A., Malik, M.I., Shehzad, F. & Khan, A.U.K. (2017) Effect of temperature on the photovoltaic performance and stability of solid-state dye-sensitized solar cells: a review. *Renewable and Sustainable Energy Reviews* 79, 946–959. doi: 10.1016/j.rser.2017.05.114.

Nawrocka, A., Zdyb, A. & Krawczyk, S. (2009) Stark spectroscopy of charge-transfer transitions in catechol-sensitized TiO_2 nanoparticles. *Chemical Physics Letters* 475(4–6), 272–276. doi: 10.1016/j.cplett.2009.05.060.

Nazeeruddin, M.K., De Angelis, F., Fantacci, S., Selloni, A., Viscardi G., Liska, P., Ito, S., Takeru, B. & Gratzel, M. (2005) Combined experimental and DFT-TDDFT computational study of photoelectrochemical cell ruthenium sensitizers. *Journal of the American Chemical Society* 127(48), 16835–16847. doi: 10.1021/ja052467l.

Nazeeruddin, M.K., Kay, A., Rodicio, I., Humphry-Baker, R., Mueller, E., Liska, P., Vlachopoulos, N. & Gratzel, M. (1993) Conversion of light to electricity by cis-X2bis(2,2′-bipyridyl-4,4′-dicarboxylate)ruthenium(II) charge-transfer sensitizers (X=Cl-, Br-, I-, CN-, and SCN-) on nanocrystalline titanium dioxide electrodes. *Journal of the American Chemical Society* 115(14), 6382–6390. doi: 10.1021/ja00067a063.

O'Regan, B. & Grätzel, M. (1991) A low-cost, high-efficiency solar cell based on dye-sensitized colloidal TiO_2 films. *Nature* 353, 737–740. doi: 10.1038/353737a0.

Omar, A., Ali, M.S. & Rahim, N.A. (2020) Electron transport properties analysis of titanium dioxide dye-sensitized solar cells (TiO_2-DSSCs) based natural dyes using electrochemical impedance spectroscopy concept: a review. *Solar Energy* 207, 1088–1121. doi: 10.1016/j.solener.2020.07.028.

Pradhan, S.C., Hagfeldt, A. & Soman, S. (2018) Resurgence of DSCs with copper electrolyte: a detailed investigation of interfacial charge dynamics with cobalt and iodine based electrolytes. *Journal of Materials Chemistry A* 6(44), 22204–22214. doi: 10.1039/C8TA06948D.

Richhariya, G., Kumar, A., Tekasakul, P. & Guptac, B. (2017). Natural dyes for dye sensitized solar cell: a review. *Renewable and Sustainable Energy Reviews* 69, 705–718. doi: 10.1016/j.rser.2016.11.198.

Saidi, N.M., Farhana, N.K., Ramesh, S. & Ramesh, K. (2021) Influence of different concentrations of 4-tert-butyl-pyridine in a gel polymer electrolyte towards improved performance of Dye-Sensitized Solar Cells (DSSC). *Solar Energy* 216, 111–119. doi: 10.1016/j.solener.2020.12.058.

Sakamoto, H., Igarashi, S., Uchida, M., Niume, K. & Nagai, M. (2012) Highly efficient all solid state dye-sensitized solar cells by the specific interaction of CuI with NCS groups II. Enhancement of the photovoltaic characteristics. *Organic Electronics* 13(3), 514–518. doi: 10.1016/j.orgel.2011.11.017.

Sánchez-de-Armas, R., San-Miguel, M.A., Oviedo, J. & Fdez Sanz, J. (2011) Direct vs. indirect mechanisms for electron injection in DSSC: catechol and alizarin. *Computational and Theoretical Chemistry* 975(1–3), 99–105. doi: 10.1016/j.comptc.2011.01.010.

Sharma, K., Sharma, V. & Sharma, S.S. (2018) Dye-sensitized solar cells: fundamentals and current status. *Nanoscale Research Letters* 13, 381. doi: 10.1186/s11671-018-2760-6.

Shi, D., Pootrakulchote, N., Li, R., Guo, J., Wang, Y., Zakeeruddin, S.M., Grätzel, M. & Wang, P. (2008) New efficiency records for stable dye-sensitized solar cells with low-volatility and ionic liquid electrolytes. *Journal of Physical Chemistry C* 112, 17046–17050. doi: 10.1021/jp808018h.

Tang, Y.-B., Lee, C.-S., Xu, J., Liu, Z.-T., Chen, Z.-H., He, Z., Cao, Y.-L., Yuan, G., Song, H., Chen, L., Luo, L., Cheng, H.-M., Zhang, W.-J., Bello, I. & Lee, S.-T. (2010) Incorporation of graphenes in nanostructured TiO_2 films via molecular grafting for dye-sensitized solar cell application. *ACS Nano* 4, 3482–3488. doi: 10.1021/nn100449w.

Tian, H., Zhang, J., Wang, X., Yu, T. & Zou, Z. (2011) Influence of capacitance characteristic on I–V measurement of dye-sensitized solar cells. *Measurement* 44(9), 1551–1555. doi: 10.1016/j.measurement.2011.06.003.

Tobaldi, D.M., Pullar, R.C., Durães, L., Matias, T., Seabra, M.P. & Labrincha, J.A. (2016) Truncated tetragonal bipyramidal anatase nanocrystals formed without use of capping agents from the supercritical drying of a TiO_2 sol. *Crystal Engineering Communication* 18, 164–177. doi: 10.1039/c5ce02112j.

Wang, Y., Wu, D., Fu, L.-M., Ai, X.-C., Xu, D., & Zhang, J.P. (2014) Density of state determination of two types of intra-gap traps in dye-sensitized solar cells and its influence on device performance. *Physical Chemistry Chemical Physics* 16, 11626–11632. doi: 10.1039/C4CP00779D.

Wu, M. & Ma, T. (2018) Transition metal compounds electrocatalysts for I-mediated dye-sensitized solar cells. In: Yun, S. & Hagfeldt, A. (eds.) *Counter Electrodes for Dye-Sensitized and Perovskite Solar Cells*. Wiley: Hoboken, NJ, pp. 155–176. doi: 10.1002/9783527813636.

Wu, K.-L., Ho, S.-T., Chou, C.-C., Chang, Y.-C., Pan, H.-A., Chi, Y. & Chou, P.T. (2012) Engineering of Osmium(II)-based light absorbers for dye-sensitized solar cells. *Angewandte Chemie* 124(23), 5740–5744. doi: 10.1002/ange.201200071.

Wu, J., Lan, Z., Lin, J., Huang, M., Huang, Y., Fan, L., Luo, G., Lin, Y., Xie, Y. & Wei, Y. (2017) Counter electrodes in dye-sensitized solar cells. *Chemical Society Reviews* 46, 5975–6023. doi: 10.1039/C6CS00752J.

Xi, C., Cao, Y., Cheng, Y., Wang, M., Jing, X., Zakeeruddin, S.M., Grätzel, M. & Wang, P. (2008) Tetrahydrothiophenium-based ionic liquids for high efficiency dye-sensitized solar cells. *Journal of Physical Chemistry C* 112, 11063–11067. doi: 10.1021/jp802798k.

Yahya, M., Bouziani, A., Ocak, C., Seferoğlu, Z. & Sillanpää, M. (2021) Organic/metal-organic photosensitizers for dye-sensitized solar cells (DSSC): recent developments, new trends, and future perceptions. *Dyes and Pigments* 192, 109227. doi: 10.1016/j.dyepig.2021.109227.

Yang, H., Sun, C., Qiao, S., Zhou, J., Liu, G., Smith, S.C., Cheng, H.M. & Lu, G.Q. (2008) Anatase TiO_2 single crystals with a large percentage of reactive facets. *Nature* 453, 638–641. doi: 10.1038/nature06964.

Yang, M., Kim, D., Jha, H., Lee, K., Paul, J. & Schmuki, P. (2011) Nb doping of TiO_2 nanotubes for an enhanced efficiency of dye-sensitized solar cells. *Chemical Communications* 47, 2032–2034. doi: 10.1039/C0CC04993J.

Yella, A., Lee, H.-W., Tsao, H.N., Yi, C., Chandiran, A.K., Nazeeruddin, M.K., Diau, E.W.-G., Yeh, C.-Y., Zakeeruddin, S.M. & Grätzel, M. (2011) Porphyrin-sensitized solar cells with cobalt (II/III)–based redox electrolyte exceed 12 percent efficiency. *Science* 334, 629–634. doi: 10.1126/science.1209688.

Yu, J., Low, J., Xiao, W., Zhou, P. & Jaroniec, M. (2014) Enhanced photocatalytic CO_2-reduction activity of anatase TiO_2 by coexposed {001} and {101} facets. *Journal of American Chemical Society* 136, 8839–8842. doi: 10.1021/ja5044787.

Zama, I., Martelli, C. & Gorni, G. (2017) Preparation of TiO_2 paste starting from organic colloidal suspension for semi-transparent DSSC photo-anode application. *Materials Science in Semiconductor Processing* 61, 137–144. doi: 10.1016/j.mssp.2017.01.010.

Zdyb, A. & Krawczyk, S. (2014) Adsorption and electronic states of morin on TiO$_2$ nanoparticles. *Chemical Physics* 443, 61–66. doi: 10.1016/j.chemphys.2014.08.009.

Zdyb, A. & Krawczyk, S. (2016) Characterization of adsorption and electronic excited states of quercetin on titanium dioxide nanoparticles. *Spectrochimica Acta, Part A: Molecular and Biomolecular Spectroscopy* 157, 197–203. doi: 10.1016/j.saa.2016.01.006.

Zheng, L., Bao, C., Lei, S., Wang, J., Li, F., Sun, P., Huang, N., Fang, L. & Sun, X. (2018) In situ growing CNTs encapsulating nickel compounds on Ni foils with ethanol flame method as superior counter electrodes of dye-sensitized solar cells. *Carbon* 133, 423–434. doi: 10.1016/j.carbon.2018.03.062.

Zhong, C., Gao, J., Cui, Y., Li, T. & Han, L. (2015) Coumarin-bearing triarylamine sensitizers with high molar extinction coefficient for dye-sensitized solar cells. *Journal of Power Sources* 273, 831–838. doi: 10.1016/j.jpowsour.2014.09.163.

Zhuang, M., Zheng, Y., Liu, Z., Huang, W. & Hu, X. (2015) Shape-dependent performance of TiO$_2$ nanocrystals as adsorbents for methyl orange removal. *RSC Advances* 5(17), 13200–13207. doi: 10.1039/C4RA14636K.

Chapter 4

Perovskite Solar Cells

4.1 INTRODUCTION

Both hybrid organic-inorganic and all-inorganic metal halide perovskites exhibit unique structural and optoelectronic properties such as extraordinary light absorption coefficient, tunable bandgap, high carrier mobility, long charge diffusion length and low exciton binding energy. Moreover, perovskites are low-cost materials that can be deposited by simple methods.

The photocells based on perovskites evolved from the technology of dye-sensitized solar cells; therefore, mesoporous titanium dioxide layer was a characteristic component of their structure. Over time, research has shown that perovskites, initially used as sensitizers, can also conduct charge carriers, and as a result of this observation, the titanium dioxide scaffold was eliminated and the thin-film planar structure of perovskite cells was implemented.

In this chapter, a brief history beginning with the first introduction of perovskite material to a photovoltaic cell is followed by a presentation of general characteristic of perovskites, operation rule of perovskite cells and usage of both hybrid organic-inorganic and all-inorganic metal halides. The description of architecture variants includes the overview of materials used in the role of electron transport and hole transport layers. This chapter also addresses the issue of hysteresis observed in the measurements of current-voltage characteristic of perovskite cells, the presence of lead in the perovskites used, as well as the problems with the stability that hindered perovskite cells from commercialization for many years and still remain the greatest challenge.

4.2 BRIEF HISTORY OF PEROVSKITE PHOTOCELLS

For the first time, organometal halide perovskites $CH_3NH_3PbI_3$ ($MAPbI_3$) and $CH_3NH_3PbBr_3$ ($MAPbBr_3$) were applied as sensitizers of mesoporous TiO_2 layer in dye-sensitized solar cells using liquid iodide/triiodide redox couple as the electrolyte (Kojima et al., 2009). The introduction of nanocrystalline perovskites replacing the dye in DSSCs presented by Kojima et al. resulted in strong absorption of light throughout the visible spectrum, reaching up to 800 nm upon the adsorption of the perovskite on TiO_2, and the solar-to-electrical conversion efficiency of 3.8% as well as high quantum efficiency in the case of $CH_3NH_3PbBr_3$ (Figure 4.1).

DOI: 10.1201/9781003196785-4

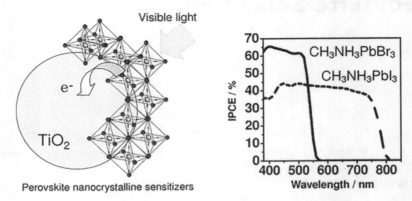

Figure 4.1 Organolead halide perovskite nanocrystals, $CH_3NH_3PbBr_3$ and $CH_3NH_3PbI_3$, sensitize TiO_2 for visible light. The $CH_3NH_3PbI_3$-based photocell exhibited spectral sensitivity of up to 800 nm, and the $CH_3NH_3PbBr_3$-based photocell demonstrated an external quantum conversion efficiency reaching 65%. (Reprinted with permission from Kojima et al. (2009). Copyright with permission from American Chemical Society.)

Then, the $CH_3NH_3PbI_3$ perovskite material in the form of quantum dots was exploited as a light harvester, and also in the liquid junction cell. The efficiency of 6.54% was achieved after optimizing the perovskite concentration and annealing temperature, with 2–3 nm sized $CH_3NH_3PbI_3$ nanocrystals deposited by spin coating of CH_3NH_3I and PbI_2 solution in γ-butyrolactone (Im et al., 2011). The core issue was the stability of the cells, since the perovskite sensitizer dissolved in the liquid electrolyte. However, this problem was overcome in further investigations in which the hole transporting material spiro-MeOTAD ([20, 7, 70-tetrakis-(N, N-di-p-methoxyphenylamine)-9,90-spiro-bifluorene]) was deployed and all-solid-state mesoscopic heterojunction cell was fabricated (Kim et al., 2012). In such a device, the photoexcitation of $CH_3NH_3PbI_3$ nanoparticles was followed by the hole transfer to spiro-MeOTAD and electron injection to submicron-thick TiO_2 film. The rapid increase in efficiency reaching 9.7% and dramatic improvement in stability was observed.

In 2012, a new type of all-solid-state dye-sensitized solar cell, based on inorganic perovskite material, was introduced. The cell was the first one in which a perovskite material was applied and worked both as a hole transport layer (HTL) (replacing the liquid electrolyte) and as the light absorber. The solar cell system consisted of the n-type nanoporous TiO_2 sensitized with the N719 dye (*cis*-diisothiocyanato-bis(2,2′-bipyridyl-4,4′-dicaboxylato) ruthenium(II) bis-(tetrabutylammonium)) and p-type semiconductor cesium tin triiodide ($CsSnI_3$) that crystalizes in three-dimensional perovskite structure (Chung et al., 2012). The inorganic perovskite $CsSnI_3$, as a solution-processable material (soluble in polar organic solvents such as acetonitrile, *N,N,*-dimethylformamide and methoxyacetonitrile), penetrated the pores of mesoporous TiO_2 layer and assured sufficient contact with dye molecules necessary to regenerate the dye after electron transfer to TiO_2. The new semiconductor introduced in DSSCs exhibited a direct bandgap of 1.3 eV, well fitted to the visible light spectrum, and a very high hole mobility of $\mu_h = 585 \text{ cm}^2/\text{Vs}$ at room temperature. Additionally, the doping of $CsSnI_3$ with F and SnF_2 significantly increased the photocurrent and the efficiency of the cells. The listed advantages of the $CsSnI_3$ semiconductor, including also the desired energies of valence band maximum, adequate for the efficient hole extraction from the dye in DSSC, resulted in the efficiency of 10.2% (Chung et al., 2012).

The 10% efficiency threshold was overcome also in meso-superstructured lead halide perovskite cell with Al_2O_3 films instead of n-type TiO_2 (Lee et al, 2012). The role of Al_2O_3 differed from that of the mesoporous TiO_2 film because the bottom of Al_2O_3 conduction band lies above the bottom of perovskite conduction band (the CB of TiO_2 lies below the CB of perovskite), which implies that there is no injection of electrons from perovskite to Al_2O_3. Therefore, the electrons remain in the perovskite layer until they are collected by anode, which is FTO-coated glass covered by a compact, planar TiO_2 film, while the Al_2O_3 layer acts only as a scaffold. In this type of a solution-processable cell, the methylammonium lead iodide chloride ($CH_3NH_3PbI_2Cl$) perovskite material transported electrons and spiro-MeOTAD transported holes. A power conversion efficiency of 10.9% and an open-circuit voltage of over 1.1 V were obtained. The observed improvement in performance occurred due to the much faster electron diffusion through the perovskite compared to the n-type TiO_2 typically used as the electron transport layer (Lee et al., 2012).

The above-mentioned examples shortly illustrate the main types of modifications introduced into DSSC-like cells in the initial evolution of perovskite solar cells, which then developed into numerous clones involving many other variations, including the use of a new type of perovskites, sometimes enriched with other materials.

4.3 OPERATION OF PEROVSKITE SOLAR CELLS

4.3.1 General characteristics of perovskite materials

Prior to the description of numerous variants of perovskite solar cells, it would be useful to gain insight into the crystalline structure and basic characteristics of this class of material. Figure 4.2 depicts the ABX_3 structure of a typical perovskite compound in which the positions of A, B and X sites are indicated. According to the widely assumed convention, in the metal halide perovskite structure the B site is a divalent metal cation (e.g., Pb and Sn) and X is a monovalent halide anion (e.g., I^- and Cl^-). In a bulk crystal structure, the A cation is surrounded by eight octahedra $[BX_6]^{4-}$ creating a cubic unit cell. The radii of a central cation are in the range of 1.6–2.5 Å.

In the solar cell, the perovskite absorber layer is typically inserted between n-type and p-type carrier-transporting materials. Upon photoexcitation of the absorber, an electron-hole exciton is generated and then split into charge carriers. The dissociation of the excitons at the perovskite/electron transport layer (ETL) interface or perovskite/hole transport layer (HTL) interface is followed by the transfer of an electron into ETL and a hole into HTL. The positions of energy levels of three main parts of PSC and the course of the charge transfer process are depicted in Figure 4.3.

The electrical conductivity in metal halide perovskites is based on the movement of electrons, holes and ions (anions and cations). Perovskite acts as an ambipolar charge conductor that shows both p-type and n-type conductivity, i.e., donor and acceptor

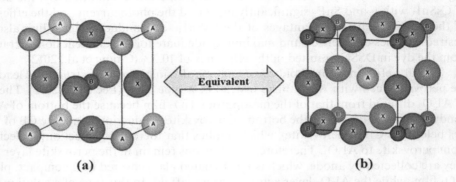

Figure 4.2 Structure of perovskite with indicated A, B and X sites. Figures (a) and (b) show the equivalent structures. (Copied under Creative Commons Attribution 3.0 License from Kumar et al. (2020).)

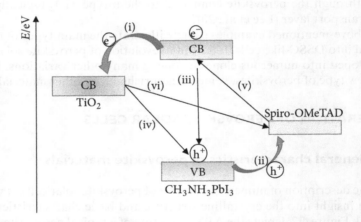

Figure 4.3 Schematic diagram of energy levels and the course of the transport of electrons and holes. The photocell consists of $CH_3NH_3PbI_3$ perovskite, TiO_2 electron transport layer and spiro-OMeTAD as a hole transport layer. (Copied under Creative Commons Attribution License from Zhou et al. (2018).)

properties. It is recognized that charge carriers can behave as large polarons, which are quasiparticles resulting from the interaction of a mobile electron or hole with the crystal lattice leading to its local deformation. However, the nature of carriers and charge transport in perovskites is still not completely clear (Batignani et al., 2018).

The perovskite materials providing the highest efficiencies of the solar cells are organic-inorganic hybrid materials. They present outstanding optoelectronic features such as a high absorption coefficient of over $1.5 \cdot 10^4 \, cm^{-1}$, long carrier lifetime, large diffusion length exceeding 1 μm as well as high mobility of electrons ($7.5 \, cm^2/V \, s$) and holes ($12.5–66 \, cm^2/V \, s$) (Grätzel, 2014; De Wolf et al., 2014). The large diffusion length of electrons and holes, which is crucial for efficient charge transport in a solar cell, reached over 100 nm in polycrystalline MAPbI$_3$ films. Moreover, it was demonstrated that in monocrystalline MAPbI$_3$, this parameter exceeded 175 μm under 1 sun and even 3 mm under weak illumination (Dong et al., 2015). Despite the numerous advantages, perovskites suffer from temperature-dependent instability. The commonly used hybrid perovskites such as methylammonium lead iodide (MAPbI$_3$) and formamidinium lead iodide (FAPbI$_3$) undergo the structure transitions upon changes in the external temperature, as shown in Figure 4.4. The ordered cubic structure of α phase (black phase) is stable at the temperature of over 330 K. Below 300 K, MAPbI$_3$ forms β phase and FAPbI$_3$ exhibits non-perovskite δ phase (yellow phase). At lower temperatures, below 160 K, MAPbI$_3$ presents orthorhombic γ phase, while below 130 K, β phase is stable for FAPbI$_3$ (Kim et al., 2020). The cubic, tetragonal and orthorhombic perovskite structures are shown in detail in Figure 4.5.

All-inorganic perovskite materials, also used in solar cells, occur in several phases as well. For instance, CsPbI$_3$, which is widely used in PSC, presents four phases, including three photoactive black phases, such as cubic α-phase, tetragonal β and

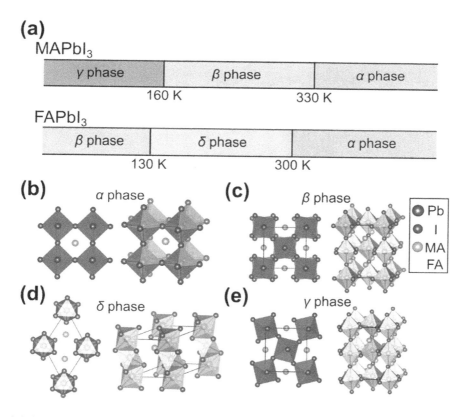

Figure 4.4 (a) The phases of MAPbI$_3$ and FAPbI$_3$ at various temperatures. The structure of (b) the cubic α phase, (c) the tetragonal β phase, (d) the trigonal δ phase and (e) the orthorhombic δ phase. (Copied under Creative Commons Attribution 4.0 International License from Kim et al. (2020).)

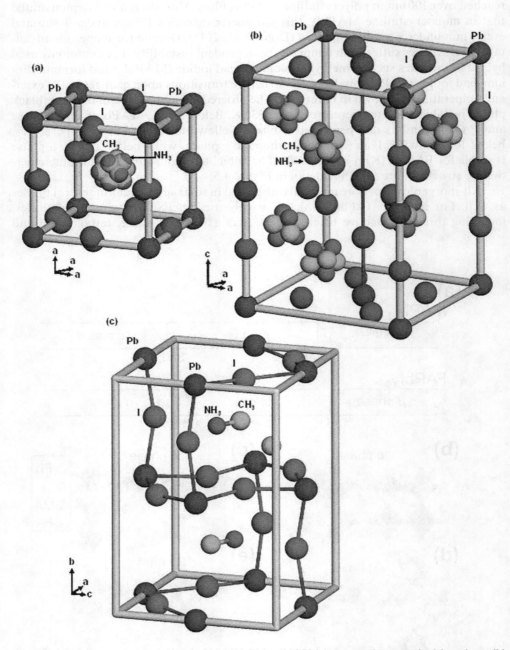

Figure 4.5 Structure models of $CH_3NH_3PbI_3$ (MAPbI$_3$) perovskite with (a) cubic, (b) tetragonal and (c) orthorhombic structures. (Copied under the terms of the Creative Commons Attribution 3.0 License from Oku (2015).)

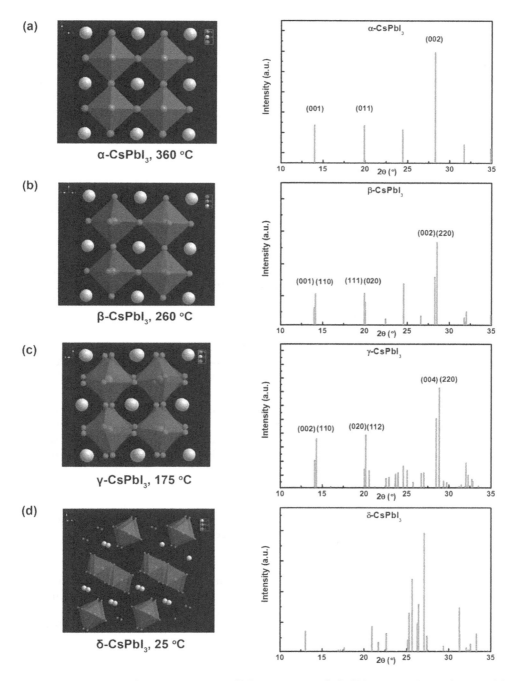

Figure 4.6 Scheme of structure and XRD patterns of CsPbI$_3$ perovskite phases. (a) α-CsPbI$_3$, (b) β-CsPbI$_3$, (c) γ-CsPbI$_3$ and (d) δ-CsPbI$_3$. (Copied with no changes under Creative Commons Attribution 4.0 International License from Supplementary Information to Wang et al. (2018a).)

orthorhombic γ, and one non-perovskite orthorhombic δ yellow phase (Figure 4.6). The phase transition from the black phase to the yellow phase at room temperature deteriorates absorption and photovoltaic performance. The γ phase transitions occur upon temperature changes. The δ phase converts to α when heated to 360°C. Upon temperature decrease, at 260°C the α phase transforms to δ, and then to γ when cooled from 175°C to RT, which is accompanied by lattice symmetry collapse. The γ phase is rather stable at RT; however, the non-photoactive δ phase prevails at RT, but it is not suitable for application in solar cells because of the large bandgap of 2.82 eV (Xiang et al., 2021).

4.3.2 Solar cells based on organic-inorganic and all-inorganic perovskites

4.3.2.1 Hybrid organic-inorganic perovskites

The most representative and widely reported in the literature are methylammonium lead iodide ($MAPbI_3$) and formamidinium lead iodide ($FAPbI_3$) hybrid organic-inorganic perovskites. Despite the promising achievements in the research on photovoltaic devices based on these perovskite materials, the problem with the reproducibility of the cells was reported due to unexpected variations in morphology. Therefore, a lot of effort has been devoted to achieving high efficiency and sufficient stability of the photocells by using the perovskite materials of desired structure obtained without compromising absorption. To this aim, different deposition methods and additional treatments were applied as well as replacements in A, B and X sites of perovskite structure were introduced.

Regarding the growth of perovskite film, it was observed that single-step deposition of $CH_3NH_3PbI_3$ using CH_3NH_3X and PbX_2 mixture resulted in uncontrolled precipitation of the material. This issue was solved by the proposed method of sequential deposition, in which the introduction of the PbI_2 solution into nanoporous TiO_2 film was followed by the exposition to CH_3NH_3I solution leading to the formation of perovskite phase (Burschka et al., 2013). The application of this method provided an efficiency of up to 15%, which permitted better control over the morphology of photoactive layer, significantly increased reproducibility and opened new routes for usage of other perovskite materials in solution-processed cells.

The method of dual-source vapor deposition was applied to obtain planar perovskite absorber, which provided a similar efficiency, over 15% (Liu et al., 2013). The fabricated cells were based on a planar heterojunction thin-film architecture made with organometal trihalide perovskite exhibiting the same crystalline structure and thickness as a solution-processed film. The comparison of the devices with vapor-deposited and solution-processed $MAPbCl_3$ showed the significant advantage of the new approach in terms of photovoltaic parameters.

Another approach to the improvement in perovskite properties, widely reported in the literature, is the tuning of absorption range, which is directly related to the bandgap value. The goal for the tailoring of bandgap, a crucial feature of the efficient absorber, was determined by the theoretical Shockley-Queisser limit, which predicts the ideal bandgap value of 1.3 eV for single-junction solar cell (Shockley & Queisser, 1961). In this respect, the most representative hybrid halide perovskite methylammonium lead iodide MAPbI$_3$ with the bandgap of 1.5–1.6 eV seems far from being optimal for efficient solar power collection (Zhu et al., 2016). However, it was experimentally demonstrated that in the cells based on MAPbI$_3$, the optimization of Ag contacts that aimed at the decrease in the sheet resistance led to a PCE of up to 18.4% (Eze et al., 2021). The typical structure of such a cell and the Tauc plot obtained to determine the bandgap value of MAPbI$_3$ are depicted in Figure 4.7.

In order to adjust the bandgap, the basic structure of perovskite can be modified by incorporation of various types of ions. In practice, one of the effective methods in bandgap tuning is mixing of different halide components (Poespawati et al., 2020). The example is MAPb(Br$_x$I$_{1-x}$)$_{3-y}$Cl$_y$ ($0 \leq x \leq 1$), of which the bandgap is linearly dependent on the percentage of Br (Iftikhar et al., 2021). The optical properties of the popular MAPbI$_3$ perovskite can be tuned by changing the halide content or mixing with other perovskite materials, which in consequence leads to the improvement in the efficiency.

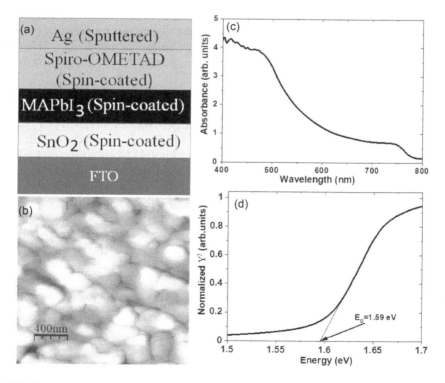

Figure 4.7 (a) The structure of perovskite cell, (b) AFM image of perovskite layer, (c) absorption spectrum of MAPbI$_3$ and (d) the Tauc plot for MAPbI$_3$ with the indicated bandgap value. (Reprinted from Eze et al. (2021). Copyright with permission from Elsevier.)

The approach based on the combination of two different photoactive perovskite materials enabled wide absorption and increased efficiency. The MAPbI$_3$ and MAPbBr$_3$ in bilayer architecture, obtained by a fully solution-based process, covered the absorption range to 770 nm and led to the efficiency of 16.2% (Jeon et al., 2014). Further replacement of MAPbI$_3$ with another widely used perovskite semiconductor formamidinium lead iodide (FAPbI$_3$) aimed at the redshift of absorption. Organolead trihalide FAPbI$_3$ exhibits a bandgap of 1.48 eV, close to the favorable value of 1.3 eV (Zhou et al., 2016); however, the drawback of perovskite containing FA$^+$ cation is the lack of stability at room temperature and transformation into non-perovskite yellow phase. Although FAPbI$_3$ performs weaker than MAPbBr$_3$ and is unstable in ambient humid atmosphere, it has a suitable bandgap of 1.48 eV and thus the combination of FAPbI$_3$ with MAPbBr$_3$ (bandgap of 2.3 eV) can exhibit a good match to the solar spectrum. The investigations showed that the gradual substitution of MA$^+$ with FA$^+$ cations in MAPbI$_3$ and MAPbBr$_3$ composition broadened the absorption range to longer wavelengths and a PCE of over 18% was achieved under STC (Jeon et al., 2015).

Additionally, the application of MA/FA mix is beneficial, since the smaller MA$^+$ cations favor crystallization of the black phase and promote its stabilization. However, the study of partial substitution of FA$^+$ with MA$^+$ or Pb$^+$ with Sn$^+$ in (MAPbI$_3$)$_{1-x}$(FASnI$_3$)$_x$ or MAPb$_{1-x}$Sn$_x$I$_3$ and FAPb$_{1-x}$Sn$_x$I$_3$ materials, aimed at obtaining a low-bandgap absorber, showed that instability problems still occur due to the volatility of MA$^+$ cations at higher temperatures (Zong et al., 2017). The temperature of 80°C is high enough to observe the release of CH$_3$NH$_2$ from MA-based perovskite films (Juarez-Perez et al., 2018).

The utilization of three different perovskites also proved to be a successful strategy. The enhancement of power conversion efficiency to over 20% was obtained by the system composed of three organic-inorganic perovskites: FAPbBr$_{3-x}$I$_x$ passivating layer on top of the (FAPbI$_3$)$_{0.85}$(MAPbBr$_3$)$_{0.15}$ film. The engineered FAPbBr$_{3-x}$I$_x$ layer acted as an electron-blocking layer at the perovskite/HTM interface and contributed to the enhancement of charge collection as well as the reduction in charge recombination. This innovative approach greatly improved the open-circuit voltage to 1.16 V and yielded a solar energy conversion efficiency of 21.3% (Cho et al., 2017).

Further promising results were provided by the exploitation of binary organic-inorganic perovskite compounds. The halide perovskite FAPbI$_3$ was used in binary alloy with inorganic CsSnI$_3$ perovskite, (FAPbI$_3$)$_{1-x}$(CsSnI$_3$)$_x$ presenting the bandgap tunable from 1.24 to 1.51 eV. The homogenous alloying of two different perovskite materials results in (FAPbI$_3$)$_{1-x}$(CsSnI$_3$)$_x$ composition in which the change of x from 0 to 1 is accompanied by shrinkage of crystal lattice and determines the bandgap value according to the following empirical formula (Zong et al., 2017):

$$E_g = -0.498x^3 + 1.412x^2 - 1.128x + 1.517. \tag{4.1}$$

The simple method of mixing the compounds in a polar solvent resulted in a clear solution that was spin-coated and annealed inducing crystallization of thin films. The adjusted composition (FAPbI$_3$)$_{0.7}$(CsSnI$_3$)$_{0.3}$ allowed achieving the bandgap of 1.3 eV ideal for application in single-junction PSC, which resulted in a PCE of up to 14.6% (Zong et al., 2017). In this study, the mean PCE of 11.9% and satisfactory repeatability of results was achieved by the devices in which the PEDOT:PSS/ITO cathode and fullerene-based anode were used.

Numerous studies devoted to the application of FA^+ based perovskites indicated that $FAPbI_3$ is a great candidate for highly efficient and stable photocells; however, reduction in defects and suppression of non-radiative recombination are required. Recently, the necessary improvement in $FaPbI_3$ quality has been achieved owing to the introduction of pseudo-halide anion $HCOO^-$ which passivated defects, improved crystallinity and, in consequence, led to the record certified 25.2% efficiency of perovskite cells (Jeong et al., 2021).

4.3.2.2 Inorganic perovskites

Although the record power conversion efficiency exceeding 25% was achieved by employing organic-inorganic hybrid perovskite, the long-term operation instability of solar cells based on hybrid perovskites is still an obstacle for their wide commercialization. In this context, the valuable alternative to hybrid perovskites involves all-inorganic perovskites that do not contain volatile organic components, as well as exhibit outstanding thermal stability and resistance against moisture. Due to the great advantages, inorganic perovskites have received much scientific attention in recent years and noteworthy progress in their photovoltaic performance is observed (Sanchez et al., 2018; Wang et al., 2019b, c).

All-inorganic perovskite solar cell was reported for the first time in 2015. The utilized cesium lead iodide ($CsPbI_3$), in which Cs atom is in the A site of perovskite structure, exhibited a favorable bandgap of 1.73 eV and yielded promising results. First, the $CsPbI_3$ applied in planar heterojunction architecture showed a photocell efficiency of 2.9% (Eperon et al., 2015); however, in the following years, the research on the modifications of $CsPbI_3$ perovskite brought much better results in terms of efficiency and stability. The improvement in the photovoltaic performance of all-inorganic perovskite cells was achieved owing to the tuning of bandgap value to the solar spectrum and stabilization of black phase structure.

The experimental studies included the formation of $CsPbI_3$ quantum dots coated by surfactant, which presented stability at ambient conditions for a long time. The adjustment of the quantum dots size allowed tailoring the bandgap, leading to a power conversion efficiency of 13.43% (Swarnkar et al., 2016, Sanehira et al., 2017). The desired value of the bandgap was also achieved by $CsPbI_3$ prepared in β phase, which after annealing at 210°C for 5 min exhibited a reduced bandgap value of 1.68 eV and a cell efficiency of 15.1% (Wang et al., 2019c).

Regarding the perovskite structure, the stabilization of black α phase of $CsPbI_3$ was provided by mixing the perovskite with sulfobetaine zwitterions (Wang et al., 2017b) or passivation by polymer polyvinylpyrrolidone (PVP) (Li et al., 2018), which resulted in an efficiency over 10% and long-term stability of the cells. Good effects including reproducible efficiency of 11.8% were also obtained by reducing the dimensions of the inorganic perovskite film with the addition of $EDAPbI_4$ (Zhang et al., 2017; Jiang et al., 2018). A greater efficiency of 14.57% was achieved by the utilization of high-quality $CsPbI_3$ black phase fabricated from a solution in a dry nitrogen environment (Wang et al., 2018b). In another approach, distorted orthorhombic black phase of CsPbI$_3$ with excellent crystallinity was obtained by the addition of hydroiodic (HI) acid and phenethylammonium iodide (PEAI) at the stage of solutions mixing and led to a solar cell efficiency of over 15% as well as very good stability (Wang et al., 2018a).

The performance of all-inorganic perovskite cell was further improved by the substitution of iodine anion in CsPbI$_3$ with other halide anions, which influences the bandgap value and the absorption of light.

The enhancement of photovoltaic performance was observed when mixed halide perovskite CsPbI$_3$Br$_{1-x}$ was used with π-conjugated 6TIC-4F employed to passivate uncoordinated defects and enhance charge transport layer. Non-radiative recombination was decreased as an effect of reduction in surface defects density through the coordination of nitrogen atoms with lead ions (Wang et al., 2020c). The beneficial role of passivation was proven by various experimental methods (XRD, wide-angle X-ray scattering and photoluminescence). Better crystallinity, increased absorbance, comparable value of the bandgap and suppression of non-radiative recombination in passivated material were revealed (Figure 4.7). The combination of experimental work with theoretical analysis turned out to be useful for the identification of the factors influencing voltage loss, such as radiative and non-radiative recombination and the thermal radiation in a photovoltaic cell in dark. The described solar device provided an efficiency of 16.1%.

Another approach with phenyltrimethylammonium bromide (PTABr) post-treatment of CsPbI$_3$ allowed obtaining gradient of Br doping and passivation of the perovskite surface, which led to a PCE of 17.06% (Wang et al., 2018c). Further improvement in efficiency was delivered by the adjustment of energy levels at the perovskite/electric contact interface and surface passivation. Broadening of the absorption spectrum and improvement in carrier lifetime led to a PCE of up to 18.4% under ambient conditions at elevated temperature (Wang et al., 2019c). The interfacial engineering, e.g., deposition of polythiophene hole acceptor on CsPbI$_2$Br, improved the efficiency to 12%–16%, owing to reduced interfacial recombination (Zeng et al., 2018; Tian et al., 2019). A similar approach using a Lewis base small molecule that passivated the CsPbI$_x$Br$_{3-x}$ perovskite surface enhanced the photostability and boosted the efficiency to over 16% through the suppression of non-radiative recombination and reduction in trap density (Wang et al., 2020c). The results of experimental characterization of CsPbI$_x$Br$_{3-x}$ perovskite film with and without passivation are shown in Figure 4.8.

In general, the Br doping of CsPbI$_3$ widens the bandgap from 1.73 eV for CsPbI$_3$ to 1.92 eV for CsPbI$_2$Br, 2.05 eV for CsPbIBr$_2$ and a large value of 2.3 eV for CsPbBr$_3$, which in consequence prevents the perovskite absorber from sufficient light harvesting (Chen & Choy, 2020). Nevertheless, various cesium lead halide perovskites are suitable for photovoltaic applications and can be applied in semitransparent cells. The advantages of the aforementioned CsPbBr$_3$ perovskite include higher mobility of charge carriers and better phase stability under ambient conditions, especially at elevated humidity and temperature compared to CsPbI$_3$ and hybrid halide perovskites.

Experimental studies showed that high-quality crystalline CsPbBr$_3$ perovskite films prepared with the addition of NH$_4$SCN exhibited improved light harvesting properties. In this material, the grain size was increased by partial substitution of halide atoms with pseudo-halide thiocyanate ions, which resulted in 8.47% PCE, a fill factor close to 80% and a high open-circuit voltage of 1.357 V (Wang et al., 2020b; Ke et al., 2016). A similar efficiency (8.57%) was achieved by the cells based on CsPbBr$_3$ doped by

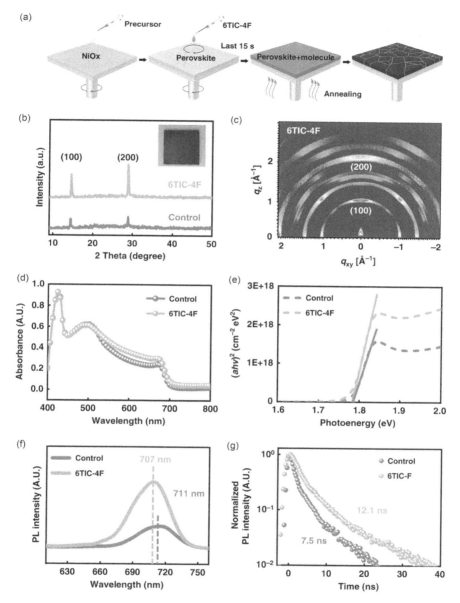

Figure 4.8 (a) Steps of a deposition process of $CsPbI_xBr_{3-x}$ film, (b) X-ray diffraction patterns of the $CsPbI_xBr_{3-x}$ film (control) and $CsPbI_xBr_{3-x}$ film with 6TIC-4F passivation (6TIC-4F), (c) grazing incidence wide-angle X-ray scattering, (d) absorption spectrum of the $CsPbI_xBr_{3-x}$ film (control) and $CsPbI_xBr_{3-x}$ film with 6TIC-4F passivation (6TIC-4F), (e) the Tauc plot, (f) steady-state photoluminescence spectra, (g) normalized time-resolved photoluminescence decay profiles And (d–g) plots for the $CsPbI_xBr_{3-x}$ film (control) and $CsPbI_xBr_{3-x}$ film with 6TIC-4F passivation (6TIC-4F). (Copied with no changes under Creative Commons Attribution 4.0 International License from Wang et al. (2020c).)

CoBr$_2$, which improved the electronic properties and prevented the formation of trap defects (Wang et al., 2020a). The improvement in material properties by reduction in halide vacancies was also obtained with guanidinium iodide (GaI) dopant in CsPbBr$_3$, and an enhancement of the efficiency to 8.54% occurred. The interesting approach in which chalcogenide quantum dots Cu, Zn, In, S and Se (CZISSe) were dispersed in the PbBr$_2$ solution and worked as nuclei seeds for the crystallization of CsPbBr$_3$ was also tested. The quantum dots promoted electron transport in the photocell from the perovskite to mesoporous TiO$_2$ layer, and as a result, the efficiency was increased by 20% of its original value (Lee et al., 2020).

The introduction of another halide anion Cl$^-$ and simple substitution of iodine anion in CsPbI$_3$ with Cl$^-$ did not provide satisfactory results, since the bandgap of over 3.0 eV characteristic of CsPbCl$_3$ is unsuitable for an efficient solar cell. However, the CsPbCl$_3$ perovskite in the form of quantum dots doped with Mn was successfully employed in photocells, which delivered a PCE of 18.57%. The quantum dots took part in the downshifting phenomenon, in which the light of short wavelength of 300–400 nm is converted into longer wavelengths of around 590 nm. The implementation of quantum dots broadened the quantum efficiency of the cells to UV region and, more importantly, prevented the degradation of perovskite absorber under UV light (Wang et al., 2017a).

The experimental investigations indicated also another role of Mn dopant in perovskite. Manganese cations were used as a partial replacement of Pb^{2+} in inorganic mixed halide perovskite CsPb$_{0.98}$Mn$_{0.02}$I$_2$Br delivering an impressive PCE of 13.47% (Bai et al., 2018). The Mn^{2+} ions passivated defects at grain boundary and surface as well as contributed to increasing the grain size and suppression of recombination. The application of other metal ions and development of CsPb$_{0.9}$Sn$_{0.1}$IBr$_2$ and CsPb$_{0.96}$Bi$_{0.4}$I$_3$ resulted in the PCE of 11.33% (Liang et al., 2017) and 13.21% (Hu et al., 2017), respectively.

Although the inorganic perovskites hold a great promise of high efficiency, their performance remains lower than perovskites with organic cations and still the issue of operational stability prevents commercialization.

4.4 ARCHITECTURE OF PEROVSKITE CELLS

Perovskite solar cells operate in several variants of architecture, which differ in the role of perovskite and electrode materials. In the traditional, mesoporous perovskite solar cell, which evolved from an n-type DSSC, the perovskite absorber is deposited onto the charge-conducting nanostructural scaffold. In the devices based on planar structure, the flat perovskite film absorbs light and, owing to its ambipolar nature, also serves as a charge conductor.

The typical mesoporous perovskite cell has the following n-i-p structure: glass electrode covered by a low work function transparent conductive oxide (TCO)/compact TiO$_2$ film/mesoporous scaffold as the electron transport layer (e.g., TiO$_2$, Al$_2$O$_3$, ZnO or SnO$_2$)/perovskite/hole transport layer (e.g., spiro-OMeTAD)/high work function counter electrode. Typically, indium tin oxide (ITO) or fluorine-doped tin oxide (FTO) constitutes the conductive coverage of illuminated electrode and the counter electrodes are made of Au or Ag; however, the degradation of metallic electrodes under

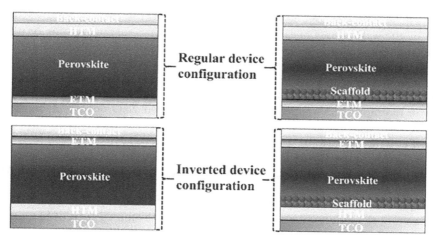

Figure 4.9 Regular configuration of PSC and inverted configuration of PSC. Planar structure (a) and scaffold-based structure (b). (Reprinted from Sajid et al. (2021). Copyright with permission from Elsevier.)

thermal stress and high price of gold prompted the search for valuable substitutions such as carbon. Electron transport layer (ETL) and hole transport layer (HTL) should form ohmic contacts with the correspondent electrodes.

The initial configuration of the perovskite cells, based on nanostructural ETL layer, was simplified by the elimination of the mesoporous layer and the introduction of planar heterojunction, which delivered a power conversion efficiency exceeding 15% (Liu et al., 2013). In such a device, the mixed halide perovskite absorber was prepared by the dual-source vapor deposition directly on the compact n-type TiO_2 film and then the p-type hole transporter spiro-OMeTAD layer was solution-processed on top of the perovskite film. The advantages of planar perovskite cells are easy processing at low temperature and low cost. The planar devices can be fabricated in the n-i-p configuration or inverted p-i-n configuration (Figure 4.9), which evolved from organic solar cells. The p-i-n architecture is obtained by the deposition of perovskite film on illuminated electrode consisting of glass covered by FTO and a hole transport layer such as PEDOT:PSS that exhibits high conductivity and sufficient transparency. In this inverted structure, the ETL is usually also an organic material, e.g., PEDOT:PSS.

4.4.1 ETL

Electron and hole transport layers in perovskite cells should efficiently extract the charges and transport them to the appropriate electrode. The basic requirements for ETL are suitable match of energy levels at the interface with perovskite and electrode, high electron mobility, easy processing and stability under ambient conditions. In the cells of n-i-p architecture, the electron transport material acts as a window layer; hence, it should exhibit a wide bandgap to transmit most of the impinging light into the perovskite absorber. The common electron transport layer is TiO_2 (bandgap 3.2 eV) fabricated by the well-known methods such as spin coating, screen printing or magnetron sputtering (Wang et al., 2019b). However, the stability of PSC with TiO_2 is not satisfactory due to the recombination processes. The substitution of TiO_2 with mesoporous Al_2O_3 resulted in an increase in the open-circuit voltage to over 1.1 V and improvement in efficiency. The valuable replacement for TiO_2 is also ZnO, characterized by higher electron mobility. The application of nanostructured ZnO deposited by the electrospraying method provided an efficiency of the cell exceeding 10% (Mahmood et al., 2014). The ZnO can also be deposited by other methods such as sol-gel, hydrothermal, chemical bath, atomic layer deposition and magnetron sputtering. Well-stabilized perovskite cells of over 16% efficiency were obtained with nitrogen-doped or Al-doped ZnO in the form of nanowires or nanorods. Another alternative to TiO_2 is SnO_2, which can be processed in low temperature, has high electron mobility and exhibits favorable alignment of the bands with perovskite material. The employment of SnO_2 nanocrystals provided a PCE close to 19%, and over 90% of initial efficiency was attained under ambient conditions (Wang et al., 2019b). Between ETL and TCO, a compact TiO_2 layer is inserted, which extracts electrons from the electron transport layer and blocks holes. The compact layer can be deposited by the sol-gel method, spray pyrolysis, atomic layer deposition or electrodeposition.

Organic materials can also be used as ETL due to their advantages, such as flexibility and easy fabrication at low temperature. The PEDOT:PSS, PCBM or fullerene derivatives are often employed as ETL, especially in the cells of p-i-n structure. Fullerenes and their derivatives are excellent materials for ETL due to high electron mobility and electron affinity as well as good alignment of energy levels with perovskites. Additionally, the application of fullerenes reduces the deep trap states present in perovskite as an effect of passivation of grain boundaries and surface trap states. An example of the structure and energy level alignment in the inverted p-i-n PSC with fullerene C_{60} is shown in Figure 4.10. Other nanocarbon materials, e.g., graphene,

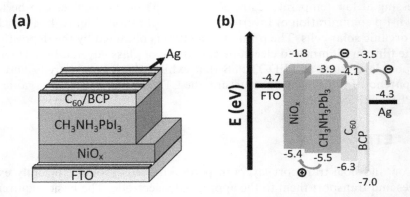

Figure 4.10 The inverted p-i-n structure configuration of perovskite cell: $FTO/NiO_x/CH_3NH_3PbI_3/C_{60}$/bathocuproine (BCP)/Ag (a); the energy level diagram of the cell (b). (Copied under Creative Commons Attribution License from Hsu et al. (2022).)

Table 4.1 The configurations and performance parameters of perovskite cells with ETL enriched by carbon

Cell configuration	J_{SC} (mA/cm^2)	V_{OC} (V)	FF (%)	Efficiency (%)	References
ITO/PEDOT:PSS/CH$_3$NH$_3$PbI$_3$/RGO-PCBM/PFN/Ag	22.92	0.85	65.8	14.5	Kakavelakis et al. (2017)
FTO/C$_{60}$/CH$_3$NH$_3$PbI$_3$/carbon	23.08	1.07	61.25	15.38	Meng et al. (2018)
TiO$_2$/bis-PCBM:DMC/perovskite/HTL/MoO$_3$/Au	23.95	1.08	77	20.14	Ye et al. (2018)
ITO/PEDOT:PSS/perovskite/PS-C$_{60}$/BCP/Al	22.9	1.1	80.6	20.3	Wang et al. (2016c)
ITO/SWCNT-TiO$_2$/perovskite/spiro-OMeTAD/Au	21.62	1.04	70	15.58	Batmunkh et al. (2017)
FTO/TiO$_2$/Li-GO/perovskite/spiro-OMeTAD/Au	19.6	0.859	70	11.14	Agresti et al. (2016)

carbon quantum dots (CQD), nanotubes (CN), graphene flakes (GF), graphene oxide (GO) and reduced graphene oxide (RGO), also find application in ETL and enhance the performance as well as the stability of perovskite solar devices due to their excellent optical and electronic properties.

The utilization of graphene nanoflakes in ETL was first demonstrated in the meso-superstructured perovskite solar cell. The introduction of low-cost, solution-processed thin-film nanocomposite of graphene and TiO$_2$ nanoparticles resulted in a decrease in series resistance, an increase in recombination resistance and a power conversion efficiency of up to 15.6% (Wang et al., 2014a). The addition of reduced graphene oxide (RGO) to PCMB electron transport layer in planar inverted perovskite cells led to a PCE of 14.5%, which means the improvement of 12% in comparison with the device with pristine PCMB (Kakavelakis et al., 2017).

In all-inorganic inverted perovskite cells, inorganic electron transport materials, e.g., MoO$_x$ or ZnO, with fullerenes provided an efficiency from 5% to over 16% (Chen & Choy, 2020). Table 4.1 depicts the various configurations and performance parameters of perovskite cells with ETL enriched by different forms of carbon.

4.4.2 HTL

The hole transport material used in the perovskite cell should efficiently transport holes, block electrons and contribute to the protection of perovskite material from external factors that deteriorate its stability. As a hole transport material, in mesoscopic perovskite photocell with n-i-p architecture, usually spiro-OMeTAD or polytriarylamine (PTAA) is applied. Spiro-MeOTAD is widely used in PSC due to the good match of energy levels, amorphous nature and high solubility. Another common HTL of choice, which is PTAA, proved to work well under illumination and high temperature, supporting the overall stability of the cell. The popular HTL is also PEDOT:PSS exhibiting high conductivity, proper alignment of energy levels, thermal stability as well as transparency and flexibility. However, the organic hole transport layer can be replaced by an inorganic p-type metal oxide in the cells of the following inverted p-i-n configuration: glass covered by TCO/HTL/perovskite/ETL/metallic electrode. The first mesoscopic p-i-n cell was based on nanocrystalline NiO and $PC_{61}BM$ that were used in the role of HTL and ETL, respectively. The PCE of 9.51% obtained by the photocell was further improved to over 11% after the introduction of sputtered NiO film (Wang, et al., 2014b).

A good solution is the usage of hole transport materials enriched with carbon nanostructures, which exhibit a higher hole extraction rate, better efficiency and stability. The improvement in hole extraction in the cell is of crucial importance, since it also positively affects the electron extraction by TiO_2 used as ETL. The rapid hole extraction (below 1 ps) accompanied by slow recombination (hundreds of microseconds) was achieved when single-walled carbon nanotubes were introduced in spiro-OMeTAD (Ihly et al., 2016). The exploitation of carbon nanotubes/graphene oxide buffer layer in the cell based on organolead iodide enhanced the efficiency and stability compared to the cell using typical HTM (Wang et al., 2016b). The CN/GO buffer layer introduced at the interface between PMMA and perovskite played a role in efficient hole transport and blocking of electrons. Another approach in which solution-processed CQD were used as all-carbon HTL demonstrated a low efficiency of 3% of the cell with methylammonium iodide perovskite. However, the improvement in all photovoltaic parameters, including a PCE of 8.06%, was achieved after the application of CQD together with spiro-OMeTAD (Paulo et al., 2016). The different configurations of the perovskite cells with carbon incorporated in HTL and their photovoltaic parameters are listed in Table 4.2.

For application in normal all-inorganic PSC, inorganic hole transport materials, which present suitable energy levels, good stability and low price, were developed. Different compounds such as CuI, Cu_2ZnSnS_4 QDs, $CuInS_2$/ZnS QDs, Cu(Cr, M)O_2 nanocrystals, Co_3O_4, NiO_x, MnS, PbS, PEI-CNT were employed as inorganic HTL;

Table 4.2 The configurations and performance parameters of perovskite cells with HTL enriched by carbon

Cell configuration	J_{SC} (mA/cm^2)	V_{OC} (V)	FF (%)	Efficiency (%)	References
FTO/TiO$_2$/perovskite/S-SWCNT-spiro-OMeTAD/Ag	22.07	1.14	75	18.9	Habisreutinger et al. (2017)
FTO/TiO$_2$/CH$_3$NH$_3$PbI$_3$/SWCNT/GO/PMMA	20.03	1.1	70	15.5	Aitola et al. (2016)
ITO/CNT-PEDOT:PSS/perovskite/PCBM/Ag	20.35	1.04	75.4	15.6	Yoon et al. (2019)
FTO/NiO/GO/perovskite/GO-Li/TiO$_x$/Al	18.6	0.97	62	14	Nouri et al. (2018)
FTO/TiO$_2$/perovskite/SWCNT-GO PMMA//Ag	17.7	0.95	60	13.3	Wang et al. (2016a)
FTO/TiO$_2$/perovskite/CQDs/Au	7.83	0.97	74	3.0	Paulo et al. (2016)

however, the obtained efficiencies were non-satisfactory, which indicated the need for the improvement in holes extraction ability (Chen & Choy, 2020).

4.4.3 ETL-free and HTL-free perovskite photocells

Numerous reports demonstrate that ETL and HTL are not necessary for the correct operation of the perovskite cells; however, simultaneous elimination of these components and maintenance of the efficiency remains a challenge. In ETL-free or HTL-free perovskite cells, one step of deposition process is eliminated and the production costs are reduced, which facilitate commercialization. The structure of such devices is simpler, and fewer materials are used.

The ETL-free perovskite cells achieve efficiencies of up to around 15%–19% for structures with Au metallic electrode and different TCO (e.g., FTO/MAPbI$_3$/spiro-OMeTAD/Au, FTO/Cs$_{0.05}$FA$_{0.81}$MA$_{0.14}$PbI$_{2.55}$Br$_{0.45}$/spiro/Au, ITO/FAPbI$_{3-x}$Cl$_x$/spiro/Au, ITO/MAPbI$_3$/spiro-OMeTAD/Au) and values of up to 15.5% with Ag metallic electrode and different TCO (e.g., ITO/MAPbI$_3$/P3HT/Ag, ITO/MAPbI$_3$/spiro-OMeTAD/Ag, FTO/MAPbI$_3$/spiro/Ag, FTO/MAPbI$_{3-x}$Cl$_x$/spiro-OMeTAD/Ag) (Huang & Ge, 2019). In the ETL-free cells, achieving the efficiency increase required the improvement in the growth process of perovskite on TCO. Severe charge accumulation due to a large electron energy barrier between TCO and perovskite in comparison with ETL-based devices with cascade energy levels is a problematic issue. Additionally, the current leakage occurs as a consequence of insufficient coverage of the substrate by perovskite film, which results in carrier recombination at the perovskite/TCO interface. Moreover, the absence of ETL means the lack of hole-blocking layer. In ETL-based cells, the optimization of ETL enables us to eliminate the hysteresis and improve the photovoltaic parameters and reproducibility.

The HTL-free perovskite cells have also been attracting the attention in recent years due to the possible lowering of costs and reduction in fabrication complexity. An interesting approach was demonstrated in HTL-free inverted PSC with ETL consisting of ZnO-RGO-CuInS$_2$ nanostructural composite and CH$_3$NH$_3$PbI$_3$ perovskite, which delivered 15.74% PCE and retained 94.7% of initial efficiency after 10 days. In such a device, the well-aligned energy levels in the ternary ETL resulted in favored cascade electron transport (Taheri-Ledari et al., 2020). Good stability and low cost can be achieved owing to the promising solution consisting in the application of carbon electrodes in HTL-free perovskite cells. The introduction of carbon material in the form of graphene, carbon nanotubes, graphite sheets or commercial carbon conductive paste, deposited by, e.g., doctor blading, screen printing or hot press transfer, enables the elimination of HTL. In order to protect perovskite from humidity in HTL-free structure of the cell, a hydrophobic carbon electrode can be introduced as a top contact. The device with the following structure: FTO/compact TiO$_2$/MAPbI$_3$/carbon electrode, in which the carbon electrode was fabricated by hot press transfer of carbon-aluminum foil, provided 6% PCE. Around 7% PCE was attained by the application of the doctor blade method followed by the 24 h of vacuum treatment, which ensured better adhesion to the underlying perovskite film (Valastro et al., 2020).

The great potential of the inverted HTL-free cells was proven by the achievement of efficiency over 20% and maintenance of 90% of the initial value under 1000 h of illumination (Zhou & Pang, 2020). The disadvantages of carbon electrodes include low hole extraction rate and non-reflective surface, which cannot work as a back mirror in the solar cell.

4.5 HYSTERESIS

Perovskite solar cells suffer from hysteresis exhibited in I-V characteristic measurements obtained with forward and reverse bias sweep. Figure 4.11 shows the difference in exemplary current-voltage characteristic registered in forward and reverse scans for the cell of the following structure: Au/spiro-OMeTAD/perovskite/SnO$_2$/FTO/glass.

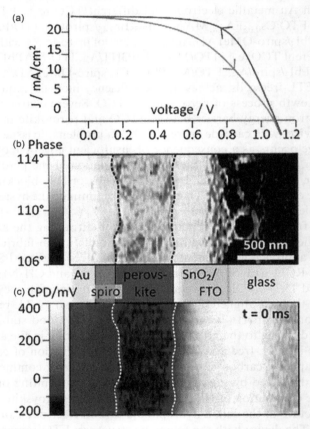

Figure 4.11 (a) Current density vs. voltage curves in forward (right arrow) and reverse (left arrow) scans; the curve was recorded after the focused ion beam treatment with a voltage sweep rate of 130 mV/s; (b) phase map; and (c) maps showing the contrast between the different layers of the solar cell. Device structure: 80–90 nm thick Au electrode, the spiro-OMeTAD hole transport layer with thickness of 80–180 nm, the perovskite layer with a varying thickness of 300–550 nm, 15 nm thick SnO$_2$/FTO. (Copied under Creative Commons Attribution 3.0 License from the Royal Society of Chemistry: Weber et al. (2018).)

Hysteresis depends on measurement parameters such as scanning rate, range, direction and history of voltage changes. The severity of hysteresis is described by the hysteresis index (Aidarkhanov et al., 2020):

$$HI = \frac{PCE_{RS} - PCE_{FS}}{PCE_{RS}}, \quad (4.2)$$

where PCE_{FS}, PCE_{RS} – the power conversion efficiency for forward and reverse scans, respectively. Due to the occurrence of hysteretic effect in perovskite cells, the photovoltaic parameters obtained from current-voltage scan present ambiguous values. Therefore, the I-V curves obtained from forward and reverse scans should be averaged or the measurement procedure should be worked out to determine the reliable efficiency values and thus avoid certification problems.

The hysteresis effect is attributed to ferroelectric features of perovskite, unbalanced charge carrier transport, migration of mobile ions and trapping of charge carriers. The analysis based on first-principle methods indicated that the ferroelectricity of halide perovskites originates mainly from charge transfer between iodine and lead atoms and from molecular cation (Kim et al., 2020). Theoretical considerations of dielectric polarization, in particular dielectric properties of halide organometal perovskites, suggest overlapping of several factors, such as lattice polarization linked to ferroelectricity, charge transport and surface effects which influence the anomalous hysteretic behavior; therefore, the analysis of these factors separately is a challenge (Wilson et al., 2019). Capacitive and non-capacitive hysteresis can be distinguished. Capacitive hysteresis occurs in cells of regular structure and originates at the ETL TiO_2/perovskite interface where charge accumulation can take place. The non-capacitive effect is independent of the device architecture; it is observed at low scan rates and is related to the influence of ionic transport on the alignment of energy levels at the interfaces or on the reactivity of the contacts (Almora et al., 2016).

Hysteresis affects the photovoltaic parameters and operational stability, which hinders commercialization of perovskite cells; therefore, a lot of effort is devoted to minimizing the hysteresis effect. In this context, it is important to optimize ETL in order to improve the electron transportation ability, reduce trap states and minimize charge accumulation. In the case of titanium dioxide, which is widely used as ETL, the suppression of recombination can be achieved by the modification of structure through the development of growth techniques or introduction of 2D nanostructures. The doping of TiO_2 with Li or Cl also leads to the decrease in trap states density and better conductivity. The replacements of TiO_2, such as ZnO and SnO_2, offer higher conductivity, excellent charge separation capability and a favorable match of energy levels. Moreover, the employment of SnO_2 leads to a reduction in interfacial charge accumulation and nearly hysteresis-free performance. The incorporation of fullerenes and their derivatives in ETL minimizes the density of trap states and facilitates charge collection, which also results in the alleviation of hysteresis (Sajid et al., 2021).

Another solution that leads to the minimization of hysteretic effect is the modification of ETL/perovskite interface through passivation of trap states by PCMB or C_{60} and the optimization of layer thickness. The application of effective passivation techniques that include multiple passivation dedicated to defects of different locations as well as the improvement in perovskite quality by the optimization of growth technique led to a PCE of 19% with negligible hysteresis (HI from 0.00 to 0.02) and good stability (Aidarkhanov et al., 2020). The achieved value of PCE and 30% improvement in lifetime of the devices was obtained owing to the introduction of multilayer ETL composed of SnO_2 quantum dots, SnO_2 nanoparticles and PMMA:PCBM accompanied with the incorporation of ethylammonium iodide (EDAI) in the mixed perovskite material.

4.6 STABILITY

The stability of ABX_3 perovskite structure is determined by the Goldschmidt tolerance factor, according to the following formula (Xiang et al., 2021):

$$t = (r_A + r_X) / \left[(r_B + r_X) \sqrt{2} \right], \tag{4.3}$$

where r_A, r_B and r_X are radii of A, B and X ions, respectively. The value of the empirical parameter t in the range of 0.8–1.1 assures the stable perovskite phase. This condition is met by hybrid organic-inorganic perovskites, as for $MAPbI_3$ this factor is 0.907 and for $FAPbI_3$ $t = 0.985$ (Xiang et al., 2021).

In the case of inorganic perovskites, the factor t, for example, for $CsPbI_3$ is equal to 0.81, which is an appropriate value, determined taking into account the radii of cesium ion 1.69 Å, lead ion 1.2 Å and iodide ion 2.16 Å. Even though $t = 0.81$ is a low value, the cesium is the best choice for site A in inorganic perovskite, since other alkali metal ions (e.g., Li^+ and Na^+) have smaller radii value and provide too low factor t. In the context of photovoltaic applications, the advantage of $CsPbI_3$ is also an appropriate bandgap value of 1.73 eV.

4.6.1 Thermal stability

The growth of the solar cell temperature due to radiative heating by the sunlight during its operation is a problem that is difficult to avoid. Real operational temperature depends on the climate conditions in a given location and can change from −40 to over 80°C. Under elevated temperature, the hybrid $MAPbI_3$ perovskite degrades to PbI_2; during this process, the MA ion decomposes and the products are released as gases. The decomposition of $MAPbI_3$ occurs above 80°C due to MA sublimation (Kim et al., 2020; Juarez-Perez et al., 2018). Gradual degradation of PSC devices is also observed at room temperature. An example is the photocell based on mixed halide perovskite $MAPbI_{3-x}Cl_x$, the performance parameters of which decrease upon aging (Figure 4.12).

Inorganic perovskites exhibit promising resistance to heat owing to the elimination of volatile organic compounds and substitution with cesium. All-inorganic perovskites usually are processed at high temperature to serve in solar cell and do not decompose up to 350°C (Xiang et al., 2021). However, changes in temperature trigger phase transitions. Unfortunately, the most useful phase in PSC – α-$CsPbI_3$ – is thermodynamically unfavorable at room temperature, in contrast to δ phase that dominates at room temperature, but is not photoactive.

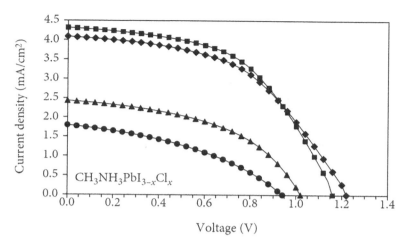

Figure 4.12 Current density-voltage characteristics of solar cell with $CH_3NH_3PbI_{3-x}Cl_x$ perovskite showing the decrease in photovoltaic parameters with time. (Copied under Creative Commons Attribution License from Poespawati et al. (2020).)

4.6.2 Illumination

In terms of the influence of illumination on the cell stability, the hybrid perovskite $MAPbBr_3$ exhibits a better resistance to photodegradation compared to $MAPbI_3$, which is assigned to comparatively stronger Pb–Br bonds than Pb–I bonds and H–Br than H–I (Misra et al., 2015). Even short-time light soaking changes the valence band and conduction band positions, which influences the bandgap value. In mixed halides, e.g., $MAPb(I_{1-x}Br_x)_3$, a manifestation of the harmful effect of solar illumination on the perovskite film is light-induced phase segregation. The phenomenon referred to as the Hoke effect is linked to facilitated migration of halide ions to the area of more stable phases and lattice strain (Slotcavage et al., 2016). The replacement of MA ions by FA or Cs at the A-site and the improvement in crystallization methods developing larger grains and limiting the vacancies can suppress the Hoke effect and thus support stability under required illumination intensity. Under illumination, the photochemical decomposition of $MAPbI_3$ occurs, while inorganic $CsPbI_3$ is more resistant to light.

The degradation, due to long-term constant illumination (light soaking), is accompanied by photocurrent decrease as the effect of formation of meta-stable trap states. Accumulation of these states leads to the formation of electric fields in macroscopic charged domains, which deteriorate the photocurrent. However, the recovery of the current in the dark was observed owing to the relaxation of light-induced trap states and dissipation of electrically charged regions in the bulk (Nie et al., 2016).

4.6.3 Moisture and oxygen

Moisture and oxygen degradation of the solar cells can be limited by proper capsulation. However, even a small amount of water or oxygen can initialize the degradation mechanism of perovskite (e.g., $MAPbI_3$), since organic ions bind water molecules, resulting in phase transformation or degradation.

In the case of inorganic perovskites, the presence of moisture enhances degradation due to the influence of oxygen and light. Water contributes to the generation of vacancies and facilitates phase transitions by lowering the energy barrier for the transition, e.g., α to δ phase transition at RT. The attempts to limit the impact of moisture include doping of perovskite material, which relaxes the lattice strains, and passivation of grain boundaries by, for example, hydrophobic additive, which in consequence reduces the number of vacancies. In inorganic perovskites, the degradation due to oxygen is significantly weaker than in organic-inorganic hybrid perovskites, and thus the inorganic perovskites containing lead can be processed in dry air. The Sn-based inorganic perovskites are sensitive to oxygen, since Sn^{+2} ion oxidizes and acts as a recombination center, which is harmful for the overall performance of the cell (Xiang et al., 2021).

4.6.4 The methods of stability improvement

One of the methods of stability enhancement is compositional engineering aiming at the development of mixed compounds. A large potential for improvement in the stability of hybrid MA-based cells was demonstrated by the partial substitution of MA with imidazolium organic cation $C_3N_2H_5^+$ (IM^+). An increase in the IPCE of around 10% and better electronic and structural properties were found based on the experimental research. The application of optimal composition $MA_{0.94}IM_{0.06}PbI_{2.6}Cl_{0.4}$ provided long-term stability exceeding 2000 h (Tomulescu et al., 2021). Another promising strategy is based on the addition of inorganic Cs to the mix of MA/FA. The usage of MA/FA/Cs cation mixture favors the formation of defect-free films by suppression of impurities resulting in a PCE of over 21% and impressive stability with 18% PCE after 250 h of operation (Saliba et al., 2016).

Due to the instability induced by labile organic cations, inorganic cations were incorporated at the A site, which led to the improvement in the stability influenced by light, humidity and temperature. The improvement in black phase stability of inorganic perovskites can be realized by the increase in factor t through the selection of appropriate replacements at B and X sites. This strategy requires testing of optical properties since each change in composition may affect the bandgap.

Various elements were employed for the B site substitution: Sn (Liang et al., 2017), Ge (Yang et al., 2018; Yang et al, 2021), Bi (Hu et al., 2017) and Mn (Bai et al., 2018). In general, the B site cations of smaller radii than Pb are preferred as replacements to increase factor t for the given material. Excellent stability was achieved with the use of $CsPbI_3$ with 5% of Ca^{2+} cations. The partial replacement of Pb by Ca passivated the surface, improved the contact between perovskite and the hole transport layer, as well as allowed achieving the efficiency of 13.3%. It was possible to maintain 85% of this efficiency for over 32 months, for the encapsulated cell (Lau et al., 2018).

The incorporation of Zn in $CsPbI_2Br$ is a good example of the multi-benefit effect. The development of $CsPb_{0.9}Zn_{0.1}I_2Br$ leads to decreases in trap states and the bandgap value as well as enhancement of photovoltaic performance and stability of the cells (Sun et al., 2019). As a result, the increase in efficiency to 13.6% compared to 11.8% for pure $CsPbI_2Br$ was obtained. The incorporation of Zn ions directly influences the phase formation process, which is illustrated in figure showing the XRD changes vs annealing temperature, grazing incidence wide-angle X-ray scattering (GIWAXS) patterns and crystallization scheme. The described examples of B site substitution in the Pb-based perovskites also imply an additional important benefit, which is the elimination of lead.

The X site substitution is a beneficial approach that can improve the factor t and thus stabilize the black phase of perovskite; however, better phase stability is accompanied by a larger bandgap, which is favorable for application in two-level tandem solar cells. The improved stability originating from Br incorporated in mixed halide perovskite $CsPbI_{2.85}Br_{0.15}$, delivering a PCE of 16.83% was demonstrated (Wang et al., 2019a). In this case, the doping of $CsPbI_3$ by 5% of Br ions led to crystalline lattice strain relaxation and reduction in recombination, and finally, thermal and compositional stability was achieved. In another approach, an optimized $CsPbI_2Br$ film obtained by the two-step deposition method followed by annealing delivered 10.21% PCE, and after storing in air for 44 days, 90% of the efficiency was retained (Dong et al., 2019).

The interesting strategy is also the replacement of halide anion by SCN, F and Ac, e.g., in $CsPbI_{2-x}Br(Ac)_x$ perovskite, which contributes to a better match of energy levels. Owing to the improved crystal quality, a PCE of 15.56% was obtained and 98% of this value was retained after 14 days in air without encapsulation (Zhao et al., 2019).

The promising method for the enhancement of perovskite phase stability is the enrichment of the precursor compound used at the deposition stage. The experimental work demonstrated a significant improvement in $CsPbI_{3-x}Br_x$ black phase stability attained after introducing $PbCl_2$ into the perovskite precursor solution. Moreover, the LiF interfacial layer inserted between glass/ITO/SnO_2 electrode and perovskite resulted in a better alignment of energy levels and passivation of defects. Due to a combination of $PbCl_2$ addition and interface modification by LiF, a high PCE of 18.64% was achieved and over 94% of this value was maintained under continuous illumination of 1 sun for 1000h (Ye et al., 2019).

An effective method of formation of a stable perovskite film is surface post-treatment. An example is $CsPbI_3$ prepared in β phase, which shows that the application of proper external treatment can lead to excellent stabilization as well as a reduced bandgap value (Wang et al., 2019c). The β-$CsPbI_3$ was passivated by choline iodine (CHI), which penetrated the cracks present on the surface, and in consequence, reduction in traps, improvement in interfaces as well as resistance to moisture and elevated temperature were observed. The encapsulated solar cell maintained 95% of efficiency under continuous illumination at the MPP for 240h in air. Good stability of about 80% was achieved for 240h under ambient conditions in the device based on lead-free tin perovskite $FASnI_3$ passivated with poly(methyl methacrylate) (PMMA) layer, which protects from oxygen and water. The passivation by PMMA also suppressed the recombination and influenced the improvement in the open-circuit voltage (Yin et al., 2020).

4.7 LEAD-FREE PEROVSKITE CELLS

The perovskites containing lead exhibit beneficial optoelectronic properties; however, the usage of lead, which is a highly toxic element, is a crucial disadvantage of these materials. Much effort is focused on the substitution of lead, as a metal harmful for the environment, with other elements such as tin, antimony, bismuth, copper and germanium. The valuable replacement is Sn, which is a group 14 metal; it has similar electronic configuration (ns^2np^2) and ionic radius (1.19 Å for Pb^{+2} and 1.18 Å for Sn^{+2}) to lead. The Sn-based perovskites are characterized by the suitable bandgap (1.3–1.4 eV) close to the theoretical Shockley-Queisser limit and the high carrier mobility.

In the early years of research, hybrid perovskites provided a PCE of 17%, while the efficiency of lead-free cells barely exceeded 6% due to the instability of Sn^{2+} oxidation state of tin (Noel et al., 2014). Further boost of PCE to 9% was achieved owing to the development of 2D/3D perovskites, showing superior crystallinity, good electronic properties and resistance to oxidation of tin (Shao et al., 2018). The devices were prepared in inverted p-i-n structure since in the n-i-p architecture, the hole extraction is difficult due to short diffusion length of holes in tin perovskites and instability induced by oxidation of Sn.

Typically, the tin-based perovskite (e.g., $FASnI_3$) spontaneously adsorbs oxygen and Sn^{2+} oxidizes to Sn^{4+}. Additionally, in such a material, the vacancy defects are formed since the Sn–I bond easily breaks, which leads to the degradation of the perovskite material. The oxidation issue and fast rate of tin perovskite crystal growth affect the quality of crystalline structure and lead to the phase transition from typical $ASnX_3$ to non-perovskite structure, which is accompanied by broadening of the bandgap up to 1.8 eV and – in consequence – limitation of the light harvesting ability (Wu et al., 2021). The undesired phase transition can be suppressed by antioxidant additives, e.g., hydrazine derivatives introduced during the deposition of perovskite films (Tai et al., 2018). The oxidation of the perovskite can also be reduced by the addition of SnF_2 or other tin halide salts to the perovskite film. Such a treatment leads to an increase in the carrier diffusion length and the improvement in carrier extraction.

Another problem that needs to be solved on the way to wider application of tin perovskites is non-radiative charge recombination induced by a large bulk defect density exceeding $10^{16} cm^{-3}$ as well as surface trap states. To this aim, the improvement in crystalline perovskite growth methods and surface passivation is necessary. Among different passivation techniques, the usage of Lewis base molecules and fabrication of 2D/3D perovskite heterojunctions are efficient ways to reduce the charge recombination.

The substitution of lead with tin in perovskites also involves the changes in energy levels alignment. The tin-based perovskite presents shallower energy levels and – in consequence – a larger band offset at the perovskite/ETL interface. However, the bandgap of the tin-based perovskites can be tuned by the changes in the composition of the typical $ASnX_3$ structure; for example, the bandgap values can vary from 1.23 eV for $MASnI_3$ to 2.4 eV for $FASnBr_3$ (Wu et al., 2021).

The lead-free perovskites address the concern of lead toxicity; however, commercialization of lead-free perovskite cells requires optimization of many features such as bandgap width, carrier diffusion length and positions of energy levels at the interfaces. The methods of improvement of lead-free perovskites include the incorporation of anti-oxidant additives in the perovskite material, application of low-dimensional structures, which reduces harmful oxidation, as well as optimization of the cell structure aiming at improvement in energy levels alignment and limitation of the interface recombination of charge carriers.

4.8 SUMMARY

The introduction of perovskites into photovoltaic cells and the use of their extraordinary optoelectronic properties, including high absorption coefficient and the ability to transport electrons and holes, led to the recognition of perovskite cells as a separate photovoltaic technology. Due to the fact that perovskite solar cells emerged from dye-sensitized solar cells, a mesoporous scaffold was used in their initial architecture; however, further investigations, including the finding that the perovskite can serve not only as light absorber, but also as charge conductor, showed that the perovskite photocells operate successfully also in a simple planar thin-film structure. Since the first use of perovskites in dye cells as sensitizers, there has been outstanding progress in the perovskite cells technology in a relatively short time. Intensive efforts in many research centers around the world have led to achieving the efficiency of single perovskite cells comparable to the mature technology of silicon photocells, which motivates to work on the improvement in the stability of perovskite cells aimed at introducing them to the market.

REFERENCES

Agresti, A., Pescetelli, S., Cinà, L., Konios, D., Kakavelakis, G., Kymakis, E. & Carlo, A.D. (2016) Efficiency and stability enhancement in perovskite solar cells by inserting lithium-neutralized graphene oxide as electron transporting layer. *Advanced Functional Materials* 26(16), 2686–2694. doi: 10.1002/adfm.201504949.

Aidarkhanov, D., Ren, Z., Lim, C.-K., Yelzhanova, Z., Nigmetova, G., Taltanova, G., Baptayev, B., Liu, F., Cheung, S.H., Balanay, M., Baumuratov, A., Djurišić, A.B., So, S.K., Surya, C., Prasad, P.N. & Ng, A. (2020) Passivation engineering for hysteresis-free mixed perovskite solar cells. *Solar Energy Materials and Solar Cells* 215, 110648. doi: 10.1016/j.solmat.2020.110648.

Aitola, K., Sveinbjornsson, K., Correa-Baena, J.-P., Kaskela, A., Abate, A., Tian, Y., Johansson, E.M.J., Grätzel, M., Kauppinen, E.I., Hagfeldt, A. & Boschloo, G. (2016) Carbon nanotube-based hybrid hole-transporting material and selective contact for high efficiency perovskite solar cells. *Energy and Environmental Science* 9, 461–466. doi: 10.1039/C5EE03394B.

Almora, O., Aranda, C., Zarazua, I., Guerrero, A. & Garcia-Belmonte, G. (2016) Noncapacitive hysteresis in Perovskite solar cells at room temperature. *ACS Energy Letters* 1(1), 209–215. doi: 10.1021/acsenergylett.6b00116.

Bai, D., Zhang, J., Jin, Z., Bian, H., Wang, K., Wang, H., Liang, L., Wang, Q. & Liu, S.F. (2018) Interstitial Mn^{2+}-driven high-aspect-ratio grain growth for low-trap-density microcrystalline films for record efficiency $CsPbI_2Br$ solar cells. *ACS Energy Letters* 3(4), 970–978. doi: 10.1021/acsenergylett.8b00270.

Batignani, G., Fumero, G., Srimath Kandada, A.R., Cerullo, G., Gandini, M., Ferrante, C., Petrozza, A. & Scopingo, T. (2018) Probing femtosecond lattice displacement upon photo-carrier generation in lead halide perovskite. *Nature Communications* 9, 1971. doi: 10.1038/s41467-018-04367-6.

Batmunkh, M., Shearer, C.J., Bat-Erdene, M., Biggs, M.J. & Shapter, J.G. (2017) Singlewalled carbon nanotubes enhance the efficiency and stability of mesoscopic perovskite solar cells. *ACS Applied Materials and Interfaces* 9(23), 19945–19954. doi: 10.1021/acsami.7b04894.

Burschka, J., Pellet, N., Moon, S.J., Humphry-Baker, R., Gao, P., Nazeeruddin, M.K. & Gratzel, M. (2013) Sequential deposition as a route to high-performance perovskite-sensitized solar cells. *Nature* 499, 316–319. doi: 10.1038/nature12340.

Chen, J. & Choy, W.C.H. (2020) Efficient and stable all-inorganic perovskite solar cells. *RRL Solar* 4(11), 2000408. doi: 10.1002/solr.202000408.

Cho, K.T., Paek, S., Grancini, G., Roldán-Carmona, C., Gao, P., Lee, Y. & Nazeeruddin, M.K. (2017) Highly efficient perovskite solar cells with a compositionally engineered perovskite/hole transporting material interface. *Energy Environmental Science* 10, 621–627. doi: 10.1039/C6EE03182J.

Chung, I., Lee, B., He, J., Chang, R.P.H. & Kanatzidis, M.G. (2012) All-solid-state dye-sensitized solar cells with high efficiency. *Nature* 485, 486–489. doi: 10.1038/nature11067.

De Wolf, S., Holovsky, J., Moon, S.-J., Löper, P., Niesen, B., Ledinsky, M., Haug, F.-J., Yum, J.-H. & Ballif, C. (2014) Organometallic halide perovskites: sharp optical absorption edge and its relation to photovoltaic performance. *The Journal of Physical Chemistry Letters* 5(6), 1035–1039. doi: 10.1021/jz500279b.

Dong, Q., Fang, Y., Shao, Y., Mulligan, P., Qiu, J., Cao, L. & Huang, J. (2015) Electron-hole diffusion lengths > 175 μm in solution-grown $CH_3NH_3PbI_3$ single crystals. *Science* 347(6225), 967–970. doi: 10.1126/science.aaa5760.

Dong, C., Han, X., Li, W., Qiu, Q. & Jinqing Wang, J. (2019) Anti-solvent assisted multi-step deposition for efficient and stable carbon-based $CsPbI_2Br$ all-inorganic perovskite solar cell. *Nano Energy* 59, 553–559. doi: 10.1016/j.nanoen.2019.02.075.

Eperon, G.E., Paternò, G.M., Sutton, R.J., Zampetti, A., Haghighirad, A.A., Cacialli, F. & Snaith, H.J. (2015) Inorganic caesium lead iodide perovskite solar cells. *Journal of Materials Chemistry A* 3, 19688–19695. doi: 10.1039/C5TA06398A.

Eze, M.C., Ugwuanyi, G., Li, M., Eze, H.U., Rodriguez, G.M., Evans, A., Rocha, V.G., Li, Z. & Min, G. (2021) Optimum silver contact sputtering parameters for efficient perovskite solar cell fabrication. *Solar Energy Materials and Solar Cells* 230, 111185. doi: 10.1016/j.solmat.2021.111185.

Grätzel, M. (2014) The light and shade of perovskite solar cells. *Nature Materials* 13, 838–842. doi: 10.1038/nmat4065.

Habisreutinger, S.N., Wenger, B., Snaith, H.J. & Nicholas, R.J. (2017) Dopant-free planar n-i-p perovskite solar cells with steady-state efficiencies exceeding 18%. *ACS Energy Letters* 2, 622–628. doi: 10.1021/acsenergylett.7b00028.

Hsu, C.-C., Yu, S.-M., Lee, K.-M., Lin, C.-J., Liou, B.-Y., & Chen, F.-R. (2022) Oxidized nickel to prepare an inorganic hole transport layer for high-efficiency and stability of $CH_3NH_3PbI_3$ perovskite solar cells. *Energies* 15, 919. doi: 10.3390/en15030919.

Hu, Y., Bai, F., Liu, X., Ji, Q., Miao, X., Qiu, T. & Zhang, S. (2017) Bismuth incorporation stabilized α-$CsPbI_3$ for fully inorganic perovskite solar cells. *ACS Energy Letters* 2, 2219–2227. doi: 10.1021/acsenergylett.7b00508.

Huang, L. & Ziyi Ge, Z. (2019) Simple, robust, and going more efficient: recent advance on electron transport layer-free perovskite solar cells. *Advanced Energy Materials* 9(24), 1900248. doi: 10.1002/aenm.201900248.

Iftikhar, F.J., Wali, Q., Yang, S., Iqbal, Y., Jose, R., Munir, S., Gondal, I.A., Khan, M.E. (2021) Structural and optoelectronic properties of hybrid halide perovskites for solar cells. *Organic Electronics* 91, 106077, 1566–1199. doi: 10.1016/j.orgel.2021.106077.

Ihly, R., Dowgiallo, A.-M., Yang, M., Schulz, P., Stanton, N.J., Reid, O.G., Ferguson, A.J., Zhu, K., Berry, J.J. & Blackburn, J.L. (2016) Efficient charge extraction and slow recombination in organiceinorganic perovskites capped with semiconducting single-walled carbon nanotubes. *Energy and Environmental Science* 9(4), 1439–1449. doi: 10.1039/C5EE03806E.

Im, J.-H., Lee, C.-R., Lee, J.-W., Park, S.-W. & Park, N.-G. (2011) 6.5% efficient perovskite quantum-dot-sensitized solar cell. *Nanoscale* 3(10), 4088–4093. doi: 10.1039/C1NR10867K.

Jeon, N., Noh, J., Kim, Y., Yang, W.S., Ryu, S. & Seok, S.I. (2014) Solvent engineering for high-performance inorganic–organic hybrid perovskite solar cells. *Nature Materials* 13, 897–903. doi: 10.1038/nmat4014.

Jeon, N., Noh, J., Yang, W., Kim, Y.C., Ryu, S., Seo, J. & Seok, S.I. (2015) Compositional engineering of perovskite materials for high-performance solar cells. *Nature* 517, 476–480. doi: 10.1038/nature14133.

Jeong, J., Kim, M., Seo, J., Lu, H., Ahlawat, P., Mishra, A., Yang, Y., Hope, M.A., Eickemeyer, F.T., Kim, M., Yoon, Y.J., Choi, I.W., Darwich, B.P., Choi, S.J., Jo, Y., Lee, J.H., Walker, B., Zakeeruddin, S.M., Emsley, L., Rothlisberger, U., Hagfeldt, A., Kim, D.S., Grätzel, M. & Kim, J.Y. (2021) Pseudo-halide anion engineering for α-FAPbI$_3$ perovskite solar cells. *Nature* 592, 381–385. doi: 10.1038/s41586-021-03406-5.

Jiang, Y., Yuan, J., Ni, Y., Yang, J., Wang, Y., Jiu, T., Yuan, M. & Chen, J. (2018) Reduced-dimensional α-CsPbX$_3$ perovskites for efficient and stable photovoltaics. *Joule* 2(7), 1356–1368. doi: 10.1016/j.joule.2018.05.004.

Juarez-Perez, E.J., Ono, L.K., Maeda, M., Jiang, Y., Hawash, Z. & Qi, Y. (2018) Photodecomposition and thermal decomposition in methylammonium halide lead perovskites and inferred design principles to increase photovoltaic device stability. *Journal of Materials Chemistry A* 6(20), 9604–9612. doi: 10.1039/C8TA03501F.

Kakavelakis, G., Maksudov, T., Konios, D., Paradisanos, I., Kioseoglou, G., Stratakis, E. & Kymakis, E. (2017) Efficient and highly air stable planar inverted perovskite solar cells with RGO oxide doped PCBM electron transporting layer. *Advanced Energy Materials* 7(7), 1602120. doi: 10.1002/aenm.201602120.

Ke, W., Xiao, C., Wang, C., Saparov, B., Duan, H.S., Zhao, D. & Yan, Y. (2016) Employing lead thiocyanate additive to reduce the hysteresis and boost the fill factor of planar perovskite solar cells. *Advanced Materials* 26, 5214–5221. doi: 10.1002/adma.201600594.

Kim, H.-S., Lee, C.-R., Im, J.-H., Lee, K.-B., Moehl, T., Marchioro, A., Moon, S.-J., Humphry-Baker, R., Yum, J.-H., Moser, J.E., Gratzel, M. & Park, N.-G. (2012) Lead iodide perovskite sensitized all-solid-state submicron thin film mesoscopic solar cell with efficiency exceeding 9%. *Scientific Reports* 2, 591. doi: 10.1038/srep00591.

Kim, B., Kim, J. & Park, N. (2020) First-principles identification of the charge-shifting mechanism and ferroelectricity in hybrid halide perovskites. *Scientific Reports* 10, 19635. doi: 10.1038/s41598-020-76742-7.

Kojima, A., Teshima, K., Shirai, Y. & Miyasaka, T. (2009) Organometal halide perovskites as visible-light sensitizers for photovoltaic cells. *Journal of American Chemical Society* 131(17), 6050–6051. doi: 10.1021/ja809598r.

Kumar, D., Yadav, R.S., Monika, M., Singh, A.K. & Rai, S.B. (2020) Synthesis techniques and applications of perovskite materials. In Tian, H. (ed.) *Perovskite Materials, Devices and Integration*. IntechOpen: London. doi: 10.5772/intechopen.86794.

Lau, C.F.J., Deng, X., Zheng, J., Kim, J., Zhang, Z., Zhang, M., Bing, J., Wilkinson, B., Hu, L., Patterson, R., Huang, S. & Ho-Baillie, A. (2018) Enhanced performance via partial lead replacement with calcium for a CsPbI$_3$ perovskite solar cell exceeding 13% power conversion efficiency. *Journal of Materials Chemistry A* 6, 5580–5586. doi: 10.1039/C7TA11154A.

Lee, M.M., Teuscher, J., Miyasaka, T., Murakami, T.N. & Snaith, H.J. (2012) Efficient hybrid solar cells based on meso-superstructured organometal halide perovskites. *Science* 338(6107), 643–647. doi: 10.1126/science.1228604.

Lee, E.J., Kim, D.H., Chang, R.P. & Hwang, D.K. (2020) Induced growth of CsPbBr$_3$ perovskite films by incorporating metal chalcogenide quantum dots in PbBr$_2$ films for performance enhancement of inorganic perovskite solar cells. *ACS Applied Energy Materials* 11, 10376–10383. doi: 10.1021/acsaem.0c01152.

Li, B., Zhang, Y., Fu, L., Yu, T., Zhou, S., Zhang, L. & Yin, L. (2018) Surface passivation engineering strategy to fully-inorganic cubic $CsPbI_3$ perovskites for high-performance solar cells. *Nature Communications* 9, 1076. doi: 10.1038/s41467-018-03169-0.

Liang, J., Zhao, P., Wang, C., Wang, Y., Hu, Y., Zhu, G., Ma, L., Liu, J. & Jin, Z. (2017) $CsPb_{0.9}Sn_{0.1}IBr_2$ based all-inorganic perovskite solar cells with exceptional efficiency and stability. *Journal of American Chemical Society* 139(40), 14009–14012. doi: 10.1021/jacs.7b07949.

Liu, M., Johnston, M. & Snaith, H. (2013) Efficient planar heterojunction perovskite solar cells by vapour deposition. *Nature* 501, 395–398. doi: 10.1038/nature12509.

Mahmood, K., Swain, B.S. & Jung, H.S. (2014) Controlling the surface nanostructure of ZnO and Al-doped ZnO thin films using electrostatic spraying for their application in 12% efficient perovskite solar cells. *Nanoscale* 6(15), 9127–9138. doi: 10.1039/C4NR02065K.

Meng, X., Zhou, J., Hou, J., Tao, X., Cheung, S.H., So, S.K. & Yang, S. (2018) Versatility of carbon enables all carbon based perovskite solar cells to achieve high efficiency and high stability. *Advanced Materials* 30(21), 1706975. doi: 10.1002/adma.201706975.

Misra, R.K., Aharon, S., Li, B., Mogilyansky, D., Visoly-Fisher, I., Etgar, L. & Katz, E.A. (2015) Temperature- and component-dependent degradation of perovskite photovoltaic materials under concentrated sunlight. *Journal of Physical Chemistry Letters* 6, 326–330. doi: 10.1021/jz502642b.

Nie, W., Blancon, J.-C., Neukirch, A.J., Appavoo, K., Tsai, H., Chhowalla, M., Alam, M., A., Sfeir, M.Y., Katan, C., Even, J., Tretiak, S., Crochet, J.J., Gupta, G. & Mohite, A.D. (2016) Light-activated photocurrent degradation and self-healing in perovskite solar cells. *Nature Communications* 7, 11574. doi: 10.1038/ncomms11574.

Noel, N.K., Stranks, S.D., Abate, A., Wehrenfennig, C., Guarnera, S., Haghighirad, A.A., Sadhanala, A., Eperon, G.E., Pathak, S.K., Johnston, M.B., Petrozza, A., Herz, M.L. & Snaith, H.J. (2014) Lead-free organic–inorganic tin halide perovskites for photovoltaic applications. *Energy and Environmental Science* 7(9), 3061–3068. doi: 10.1039/C4EE01076K.

Nouri, E., Mohammadi, M.R. & Lianos, P. (2018) Improving the stability of inverted perovskite solar cells under ambient conditions with graphene-based inorganic charge transporting layers. *Carbon* 126, 208–214. doi: 10.1016/j.carbon.2017.10.015.

Oku, T. (2015) Crystal structures of $CH_3NH_3PbI_3$ and related perovskite compounds used for solar cells. In: Kosyachenko, L.A. (ed.) *Solar Cells: New Approaches and Reviews*. IntechOpen: London. https://www.intechopen.com/chapters/48267, doi: 10.5772/59284.

Paulo, S., Stoica, G., Cambarau, W., Martinez-Ferrero, E. & Palomares, E. (2016) Carbon quantum dots as new hole transport material for perovskite solar cells. *Synthetic Metals* 222, 17–22. doi: 10.1016/j.synthmet.2016.04.025.

Poespawati, N.R., Sulistianto, J., Abuzairi, T. & Purnamaningsih, R.W. (2020) Performance and stability comparison of low-cost mixed halide perovskite solar cells: $CH_3NH_3PbI_3$-Cl and $CH_3NH_3PbI_3$-SCN_x. *International Journal of Photoenergy* 8827917. doi: 10.1155/2020/8827917.

Sajid, S., Huang, H., Ji, J., Jiang, H., Duan, M., Liu, X., Liu, B. & Li, M. (2021) Quest for robust electron transporting materials towards efficient, hysteresis-free and stable perovskite solar cells. *Renewable and Sustainable Energy Reviews* 152, 111689. doi: 10.1016/j.rser.2021.111689.

Saliba, M., Matsui, T., Seo, J.-Y., Domanski, K., Correa-Baena, J.-P., Nazeeruddin, M.K., Zakeeruddin, S.M., Tress, W., Abate, A., Hagfeldt, A. & Gratzel, M. (2016) Cesium-containing triple cation perovskite solar cells: improved stability, reproducibility and high efficiency. *Energy and Environmental Science* 9(6), 1989–1997. doi: 10.1039/C5EE03874J.

Sanchez, S., Christoph, N., Grobety, B., Phung, N., Steiner, U., Saliba, M., Abate, A. (2018) Efficient and stable inorganic perovskite solar cells manufactured by pulsed flash infrared annealing. *Advance Energy Materials* 30, 1802060. doi: 10.1002/aenm.201802060.

Sanehira, E.M., Marshall, A.R., Christians, J.A., Harvey, S.P., Ciesielski, P.N., Wheeler, L.M., Schulz, P., Lin, L.Y., Beard, M.C. & Joseph M. Luther, J.M. (2017) Enhanced mobility $CsPbI_3$

quantum dot arrays for record-efficiency, high-voltage photovoltaic cells. *Science Advances* 3(10), eaao4204. doi: 10.1126/sciadv.aao4204.

Shao, S., Liu, J., Portale, G., Fang, H.H., Blake, G.R., ten Brink, G.H., Koster, L.J.A., & Loi, M.A. (2018) Highly reproducible Sn-based hybrid perovskite solar cells with 9% efficiency. *Advanced Energy Materials* 8, 1702019. doi: 10.1002/aenm.201702019.

Shockley, W. & Queisser, H.J. (1961) Detailed balance limit of efficiency of pn junction solar cells. *Journal of Applied Physics* 32, 510. doi: 10.1063/1.1736034.

Slotcavage, D.J., Karunadasa, H.I. & McGehee, M.D. (2016) Light-induced phase segregation in halide-perovskite absorbers. *ACS Energy Letters* 1, 1199–1205. doi: 10.1021/acsenergylett.6b00495.

Sun, H.R., Zhang, J., Gan, X.L., Yu, L.T., Yuan, H.B., Shang, M.H., Lu, C.J., Hou, D.G., Hu, Z.Y., Zhu, Y.J. & Han, L.Y. (2019) Pb-Reduced $CsPb_{0.9}Zn_{0.1}I_2Br$ thin films for efficient perovskite solar cells. *Advanced Energy Materials* 9(25), 1900896. doi: 10.1002/aenm.201900896.

Swarnkar, A., Marshall, A.R., Sanehira, E.M., Chernomordik, B.D., Moore, D.T., Christians, J.A., Chakrabarti, T. & Luther, J.M. (2016) Quantum dot–induced phase stabilization of α-$CsPbI_3$ perovskite for high-efficiency photovoltaics. *Science* 354, 92–95. doi: 10.1126/science.aag2700.

Taheri-Ledari, R., Valadi, K. & Maleki, A. (2020) High-performance HTL-free perovskite solar cell: an efficient composition of ZnO NRs, RGO, and $CuInS_2$ QDs, as electron-transporting layer matrix. *Progress in Photovoltaics: Research and Applications* 28(9), 956–970. doi: 10.1002/pip.3306.

Tai, Q., Guo, X., Tang, G., You, P., Ng, T.-W., Shen, D., Cao, J., Liu, C.-K., Wang, N., Zhu, Y., Lee, C.-S. & Yan, F. (2018) Antioxidant grain passivation for air-stable tin-based perovskite solar cells. *Angewandte Chemie International Edition* 58(3), 806–810. doi: 10.1002/anie.201811539.

Tian, J., Xue, Q., Tang, X., Chen, Y., Li, N., Hu, Z., Shi, T., Wang, X., Huang, F., Brabec, C.J., Hin-Lap Yip, H.-L. & Cao, Y. (2019) Dual interfacial design for efficient $CsPbI_2Br$ perovskite solar cells with improved photostability. *Advanced Materials* 31(23), 1901152. doi: 10.1002/adma.201901152.

Tomulescu, A.G., Leonat, L.N., Neațu, F., Stancu, V., Toma, V., Derbali, S., Neațu, S., Rostas, A.M., Beșleagă, C., Pătru, R., Pintilie, I. & Florea, M. (2021) Enhancing stability of hybrid perovskite solar cells by imidazolium incorporation. *Solar Energy Materials and Solar Cells* 227, 111096 doi: 10.1016/j.solmat.2021.111096.

Valastro, S., Smecca, E., Sanzaro, S., Giannazzo, F., Deretzis, I., La Magna, A., Numata, Y., Jena, A.K., Miyasaka, T., Gagliano, A. & Alberti, A. (2020) Improved electrical and structural stability in HTL-free perovskite solar cells by vacuum curing treatment. *Energies* 13, 3953. doi: 10.3390/en13153953.

Wang, J.T.-W., Ball, J.M., Barea, E.M., Abate, A., Alexander-Webber, J.A., Huang, J., Saliba, M., Mora-Sero, I., Bisquert, J., Snaith, H.J. & Nicholas, R.J. (2014a) Low-temperature processed electron collection layers of graphene/TiO_2 nanocomposites in thin film perovskite solar cells. *Nano Letters* 14(2), 724–730. doi: 10.1021/nl403997a.

Wang, K.-C., Shen, P.-S., Li, M.-L., Chen, S., Lin, M.-W., Chen, P. & Guo, T.-F. (2014b) Low-temperature sputtered nickel oxide compact thin film as effective electron blocking Layer for mesoscopic NiO/$CH_3NH_3PbI_3$ perovskite heterojunction solar cells. *ACS Applied Materials and Interfaces* 6(15), 11851–11858. doi: 10.1021/am503610u.

Wang, F., Endo, M., Mouri, S., Miyauchi, Y., Ohno, Y., Wakamiya, A., Murata, Y. & Matsuda, K. (2016a) Highly stable perovskite solar cells with an all-carbon hole transport layer. *Nanoscale* 8(23), 11882–11888. doi: 10.1039/C6NR01152G.

Wang, J., Li, J., Xu, X., Xu, G. & Shen, H. (2016b) Enhanced photovoltaic performance with carbon nanotubes incorporating into hole transport materials for perovskite solar cells. *Journal of Electronic Materials* 45, 5127–5132. doi: 10.1007/s11664-016-4724-x.

Wang, Q., Dong, Q., Li, T., Gruverman, A. & Huang, J. (2016c) Thin insulating tunneling contacts for efficient and water-resistant perovskite solar cells. *Advanced Materials* 28, 6734–6739. doi: 10.1002/adma.201600969.

Wang, Q., Zhang, X., Jin, Z., Zhang, J., Gao, Z., Li, Y. & Liu, S.F. (2017a) Energy-down-shift $CsPbCl_3$: Mn quantum dots for boosting the efficiency and stability of perovskite solar cells. *ACS Energy Letters* 2(7), 1479–1486. doi: 10.1021/acsenergylett.7b00375.

Wang, Q., Zheng, X., Deng, Y., Zhao, J., Chen, Z. & Huang, J. (2017b) Stabilizing the α-phase of $CsPbI_3$ perovskite by sulfobetaine zwitterions in one-step spin-coating films. *Joule* 1(2), 2017, 371–382. doi: 10.1016/j.joule.2017.07.017.

Wang, K., Jin, Z., Liang, L., Bian, H., Bai, D., Wang, H., Zhang, J., Wang, Q. & Liu, S. (2018a) All-inorganic cesium lead iodide perovskite solar cells with stabilized efficiency beyond 15%. *Nature Communications* 9, 4544. doi: 10.1038/s41467-018-06915-6.

Wang, P., Zhang, X., Zhou, Y., Jiang, Q., Ye, Q., Chu, Z., Li, Q., Yang, Y., Yin, Z. & You, J. (2018b) Solvent-controlled growth of inorganic perovskite films in dry environment for efficient and stable solar cells. *Nature Communication* 9, 2225. https://www.nature.com/articles/s41467-018-04636-4.

Wang, Y., Zhang, T., Kan, M. & Zhao, Y. (2018c) Bifunctional stabilization of allinorganic α-$CsPbI_3$ perovskite for 17% efficiency photovoltaics. *Journal of American Chemical Society* 140, 12345–12348. doi: 10.1021/jacs.8b07927#.

Wang, H., Bian, H., Jin, Z., Zhang, H., Liang, L., Wen, J., Wang, Q., Ding, L. & Liu, S.F. (2019a) Cesium lead mixed-halide perovskites for low-energy loss solar cells with efficiency beyond 17%. *Chemical Materials* 31(16), 6231–6238. doi: 10.1021/acs.chemmater.9b02248.

Wang, R., Mujahid, M., Duan, Y., Wang, Z.K., Xue, J. & Yang, Y. (2019b) A review of perovskites solar cell stability. *Advanced Functional Materials* 47, 1808843. doi: 10.1002/adfm.201808843.

Wang, Y., Dar, M.I., Ono, L.K., Zhang, T., Kan, M., Li, Y., Zhang, L., Wang, X., Yang, Y., Gao, X., Qi, Y., Grätzel, M. & Zhao, Y. (2019c) Thermodynamically stabilized β-$CsPbI_3$-based perovskite solar cells with efficiencies >18. *Science* 365(6453), 591–595. doi: 10.1126/science.aav8680.

Wang, D., Li, W., Du, Z., Li, G., Sun, W., Wu, J. & Lan, Z. (2020a) $CoBr_2$-doping-induced efficiency improvement of $CsPbBr_3$ planar perovskite solar cells. *Journal of Materials Chemistry C* 5, 1649–1655. doi: 10.1039/C9TC05679C.

Wang, D., Li, W., Du, Z., Li, G., Sun, W., Wu, J. & Lan, Z. (2020b) Highly efficient $CsPbBr_3$ planar perovskite solar cells via additive engineering with NH_4SCN. *ACS Applied Materials Interfaces* 9, 10579–10587. doi: 10.1021/acsami.9b23384.

Wang, J., Zhang, J., Zhou, Y., Liu, H., Xue, Q., Li, X., Chueh, C.-C., Yip, H.-L., Zhu, Z. & Jen, A.K.Y. (2020c) Highly efficient all-inorganic perovskite solar cells with suppressed non-radiative recombination by a Lewis base. *Nature Communication* 11, 177. doi: 10.1038/s41467-019-13909-5.

Weber, S.A.L., Hermes, I.M., Turren-Cruz, S.-H., Gort, C., Bergmann, V.W., Gilson, L., Hagfeldt, A., Graetzel, M., Tress, W. & Berger, R. (2018) How the formation of interfacial charge causes hysteresis in perovskite solar cells. *Energy Environmental Science* 11, 2404–2413. doi: 10.1039/C8EE01447G.

Wilson, J.N., Frost, J.M., Wallace, S.K. & Walsh, A. (2019) Dielectric and ferroic properties of metal halide perovskites. *APL Materials* 7, 010901. doi: 10.1063/1.5079633.

Wu, T., Liu, X., Luo, X., Lin, X., Cui, D., Wang, Y., Segawa, H., Zhang, Y. & Han, L. (2021) Lead-free tin perovskite solar cells. *Joule* 5(4), 863–886. doi: 10.1016/j.joule.2021.03.001.

Xiang, W., Liu, S.F. & Tress, W. (2021) A review on the stability of inorganic metal halide perovskites: challenges and opportunities for stable solar cells. *Energy Environmental Science* 14, 2090. doi: 10.1039/d1ee00157d.

Yang, F., Hirotani, D., Kapil, G., Kamarudin, M.A., Ng, C.H., Zhang, Y., Shen, Q. & Hayase, S. (2018) All-Inorganic $CsPb_{1-x}Ge_xI_2Br$ perovskite with enhanced phase stability and photovoltaic performance. *Angewandte Chemistry International Edition* 57(39), 12745–12749. doi: 10.1002/anie.201807270.

Yang, D., Zhang, G., Lai, R., Cheng, Y., Lian, Y., Rao, M., Huo, D., Lan, D., Zhao, B. & Di, D. (2021) Germanium-lead perovskite light-emitting diodes. *Nature Communication* 12, 4295. doi: 10.1038/s41467-021-24616-5.

Ye, Q.-Q., Wang, Z.-K., Li, M., Zhang, C.-C., Hu, K.-H. & Liao, L.-S. (2018) N-type doping of fullerenes for planar perovskite solar cells. *ACS Energy Letters* 3(4), 875–882. doi: 10.1021/acsenergylett.8b00217.

Ye, Q., Zhao, Y., Mu, S., Ma, F., Gao, F., Chu, Z., Yin, Z., Gao, P., Zhang, X. & You, J. (2019) Cesium lead inorganic solar cell with efficiency beyond 18% via reduced charge recombination. *Advanced Materials* 33(10), 1905143. doi: 10.1002/adma.201905143.

Yin, Y., Wang, M., Malgras, V. & Yamauchi, Y. (2020) Stable and efficient tin-based perovskite solar cell via semiconducting–insulating structure. *ACS Applied Energy Materials* 3(11), 10447–10452. doi: 10.1021/acsaem.0c01422.

Yoon, S., Ha, S.R., Moon, T., Jeong, S.M., Ha, T.-J., Choi, H. & Kang, D.-W. (2019) Carbon nanotubes embedded poly(3,4-ethylenedioxythiophene): poly(- styrenesulfonate) hybrid hole collector for inverted planar perovskite solar cells. *Journal of Power Sources* 435, 226765. doi: 10.1016/j.jpowsour.2019.226765.

Zeng, Q., Zhang, X., Feng, X., Lu, S., Chen, Z., Yong, X., Redfern, S.A.T., Wei, H., Wang, H., Shen, H., Zhang, W., Zheng, W., Zhang, H., Tse, J.S. & Yang, B. (2018) Polymer-passivated inorganic cesium lead mixed-halide perovskites for stable and efficient solar cells with high open-circuit voltage over 1.3 v. *Advanced Materials* 30(9), 1705393. doi: 10.1002/adma.201705393.

Zhang, T., Dar, M.I., Li, G., Xu, F., Guo, N., Grätzel, M. & Zhao, Y. (2017) Bication lead iodide 2d perovskite component to stabilize inorganic α-CsPbI$_3$ perovskite phase for high-efficiency solar cells. *Science Advances* 3(9), e1700841. doi: 10.1126/sciadv.1700841.

Zhao, H., Han, Y., Xu, Z., Duan, C., Yang, S., Yuan, S., Yang, Z., Liu, Z. & Liu, S.F. (2019) A novel anion doping for stable CsPbI$_2$Br perovskite solar cells with an efficiency of 15.56% and an open circuit voltage of 1.30 V. *Advanced Energy Materials* 9(40), 1902279. doi: 10.1002/aenm.201902279.

Zhou, Z. & Pang, S. (2020) Highly efficient inverted hole-transport-layer-free perovskite solar cells. *Journal of Materials Chemistry A* 8, 503–512. doi: 10.1039/C9TA10694D.

Zhou, Y., Yang, M., Pang, S., Zhu, K. & Padture, N.P. (2016) Exceptional morphology-preserving evolution of formamidinium lead triiodide perovskite thin films via organic-cation displacement. *Journal of American Chemical Society* 138(17), 5535–5538. doi: 10.1021/jacs.6b02787.

Zhou, D., Zhou, T., Tian, Y., Zhu, X. & Tu, Y. (2018) Perovskite-based solar cells: materials, methods, and future perspectives. *Journal of Nanomaterials*. doi: 10.1155/2018/8148072.

Zhu, Q., Bao, X., Yu, J., Zhu, D., Qiu, M., Yang, R. & Dong, L. (2016) Compact layer free perovskite solar cells with a high-mobility hole-transporting layer. *ACS Applied Materials Interfaces* 8, 4, 2652–2657. doi: 10.1021/acsami.5b10555.

Zong, Y., Wang, N., Zhang, L., Ju, M.-G., Zeng, X.C., Sun, X.W., Zhou, Y. & Padture, N.P. (2017) Homogenous alloys of formamidinium lead triiodide and cesium tin triiodide for efficient ideal-bandgap perovskite solar cells. *Angewandte Chemie International Edition* 56(41), 12658–12662. doi: 10.1002/anie.201705965.

Chapter 5

Quantum Dot-sensitized Solar Cells

5.1 INTRODUCTION

The promising photovoltaic technology of quantum dot solar cells was developed on the basis of dye-sensitized solar cells after the substitution of dye molecules with quantum dots in order to fabricate more efficient and stable devices. The distinct optoelectronic properties of quantum dots, such as high absorption coefficient, tunable range of absorption and facile fabrication at low cost, make them promising a light harvesting material for next-generation photovoltaics. Moreover, the phenomenon of carrier multiplication observed in quantum dots can influence the increase in the photocurrent. This effect, called multiple exciton generation, allows two or more excitons to be generated as a result of absorption of a single high-energy photon.

This chapter covers the basic architecture and operation cycle of quantum dot-sensitized solar cells (QDSSC), the properties of quantum dots including their improvements achieved by doping, alloying, fabrication of core-shell structure and co-sensitization method, description of multiple exciton generation, as well as in situ and ex situ deposition methods of quantum dots. The materials composing other parts of the photocells are also presented, such as carbon nanostructures and alternative electrolytes.

5.2 STRUCTURE AND OPERATION PRINCIPLE OF QUANTUM DOT-SENSITIZED SOLAR CELLS

The basic architecture of quantum dot solar cells is the same as dye-sensitized solar cells; however, dye sensitizers in these devices are replaced by semiconductor nanoparticles, i.e., quantum dots. The incident photons of energy higher than the bandgap of quantum dots are absorbed by quantum dots. The excitons created in this way dissociate into electrons transferred to electron transport layer (ETL) and the holes to hole transport layer (HTL). In QDSSC, the electron transport layer is a nanostructured wide-bandgap semiconductor, which offers large surface area, necessary to adsorb sufficient amount of quantum dots (Kouhnavard et al., 2014; Chung et al., 2021). The quantum dots (QDs) adsorbed on the surface of the semiconductor should be capable of injecting electrons into its conduction band. After the injection of the electron into the conduction band of ETL, which is typically nanostructured TiO_2, the original state of QD is restored by the redox system of the electrolyte. The injected electrons

DOI: 10.1201/9781003196785-5

migrate through the TiO$_2$ and then through the external load to the counter electrode, where the electrolyte containing the redox system is regenerated. For efficient charge transfer, the bottom of QD conduction band should be at least 0.25 eV higher than the bottom of TiO$_2$ conduction band and the top edge of QD valence band should be lower than the redox level of the electrolyte. Figure 5.1 shows the schematic structure and operating principle of QDSSC. The transport of charge carriers in the desired direction is accompanied by the recombination and relaxation processes, which hamper the

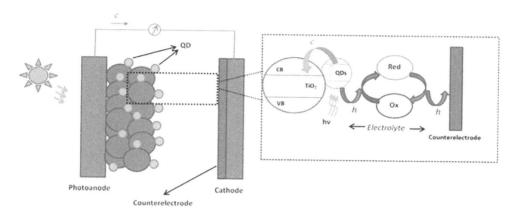

Figure 5.1 Scheme of structure and operation principle of the quantum dot-sensitized solar cell. (Reprinted from Kouhnavard et al. (2014). Copyright with permission from Elsevier.)

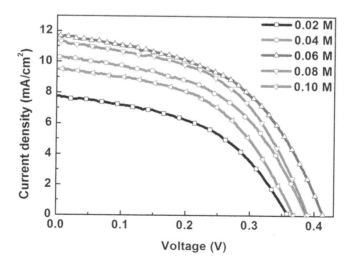

Figure 5.2 Current density vs. voltage characteristics of QDSSC with different amounts of PbS quantum dots. The load of PbS depends on the concentration of QDs precursor solution, which ranges from 0.02 to 0.1 M. (Copied under Creative Commons Attribution 4.0 International License from Tian et al. (2016).)

overall performance of the photocell. Recombination losses occur between electrons from the conduction band of sensitizing QD or TiO_2 nanoparticles and holes present in the valence band of sensitizing QD or in electrolyte.

For the quantum dots to serve well in the solar cell, their concentration on TiO_2 has to be optimized, which is illustrated in Figure 5.2. As Figure 5.2 shows, the photovoltaic parameters such as current, voltage, power and finally the efficiency generated by the QDSSC increase the concentration of Pb QDs precursor solution by up to 0.06 M, and then at higher concentrations, these parameters uniformly decrease, which is attributed to the intensification of surface charge recombination and drop in electron collection (Tian et al., 2016). Recently, the secondary deposition method has proved to be a good method for the increase in QDs load without the introduction of new recombination centers (Song et al., 2021). Due to higher loading of Zn-Cu-In-S-Se quantum dots on photoanode, a certified power conversion efficiency record of 15.2% was obtained by the cells with liquid electrolyte.

5.3 QUANTUM DOTS AS SENSITIZERS

Quantum dots are nanocrystals composed of elements from II-VI, III-V or IV-VI groups of periodic table (e.g., CdS, CdSe, CdTe and PbS) (Arivarasan et al., 2020; Huang et al., 2016). They can be exploited as sensitizers in solar cells owing to their superior optoelectronic features, which are dependent of both size and shape. Typically, the diameters of quantum dots are less than 10 nm (Sahu et al., 2020); hence, according to the laws of quantum mechanics, the dots exhibit properties in between a bulk material and a single atom or molecule. Quantum dots are a valuable substitution for dyes in the role of sensitizer in a solar cell, since they enable absorption of light in a broad range due to tunable bandgap and high absorption coefficient, as well as possess advantages such as easy deposition from solution and photostability. Additionally, quantum dots show the exceptional ability of utilizing the excess energy of absorbed photons if their energy is larger than the bandgap of QD. This phenomenon, referred to as multiple exciton generation (MEG), potentially increases the photocurrent and the efficiency of the solar cell (v.i.).

The crucial features of an absorber material in the solar cell are the proper bandgap value and the desired alignment of band edges at the interfaces of photoactive layer with ETL and HTL. In this context, the essential property offered by quantum dots is tailored QDs size and composition, which allow controlling the bandgap width and positions of energy bands. The tunable bandgap of QDs arises from the quantum size effects occurring in nanostructures. A wider bandgap is exhibited by the quantum dots of smaller size, which in consequence demonstrates the blueshift of absorption peak, depicted in Figure 5.3.

If the light absorption range of quantum dots does not cover the whole visible to near-infrared spectrum, the positions of energy bands and the bandgap width of sensitizing QDs can be modified by the following methods: doping of QDs with other elements, alloying of different materials, fabrication of core-shell nanostructures or mixing of different types of QDs. Each of these methods can also contribute to the improvement in other electronic features of quantum dots.

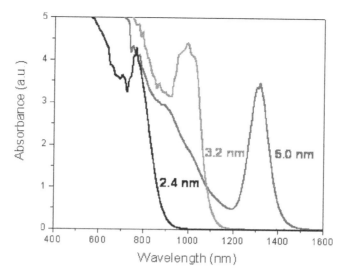

Figure 5.3 Absorption spectra of PbS quantum dots of different sizes. (Copied under Creative Commons Attribution 3.0 License from Jasim (2015).)

5.3.1 Doped quantum dots

The CdS dots applied as a sensitizer in the ZnO wide-bandgap electron transport layer are widely used in QDSSC. The doping of CdS with transition metal ions leads to the broadening of the bandgap, as well as suppression of recombination and enhancement in electron conduction. The Mn-doped CdS dots prepared by the chemical method on polyvinyl alcohol matrix with 2% Mn concentration provided a power conversion efficiency (PCE) of 2.09% and FF = 67%. This approach resulted in the enhancement of performance in comparison with a PCE of 0.99% and FF of 55% for undoped CdS quantum dots (Ganguly & Nath, 2020).

The usage of Mn-doped CdS dots in co-sensitization with CdSe dots on TiO_2 also brought much better photovoltaic parameters of the QDSSC. The obtained redshift of absorption spectra to around 650 nm due to doping with Mn and the improvement in the IPCE of up to 68% led to a PCE of 7.16% (Li et al., 2018). The experimental study demonstrated that Mn^{2+} doping of quantum dots broadens the absorption range, improves the light harvesting efficiency and decreases the electron-hole recombination within the QD. The relative positions of energy levels in a CdS nanoparticle, Mn atom and TiO_2 ETL are shown in Figure 5.4.

Rare earth elements, known for their own exclusive optical properties, also proved to be a promising dopant for QDs. The Gd doping with concentration from 0% to 20% and change in CdTe nanoparticle size from 1.8 to 3.9 nm was accompanied by the redshift of absorption maximum (from 470 to 492 nm) and a decrease in the bandgap from 2.64 to 2.52 eV. The increase in the photoconversion efficiency of up to 2.24%

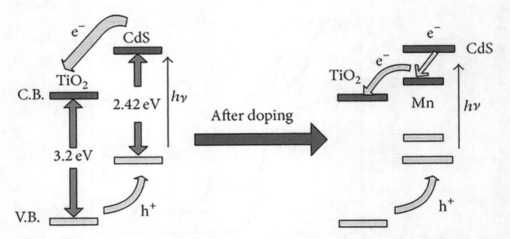

Figure 5.4 The diagram of energy levels for CdS quantum dot adsorbed on TiO_2, before and after doping with Mn. (Copied under Creative Commons Attribution License from Tianxing et al. (2014).)

compared to 0.48% for pristine CdTe was observed for 10% concentration of Gd as the dopant in CdTe QDs (Arivarasan et al., 2020).

5.3.2 Alloy dots

As an alternative to two-component quantum dots, the introduction of ternary or quaternary alloy dots in QDSSC is a promising solution due to the possible broadening of absorption range. The example includes the ternary CdSeTe QDs, which provided a PCE of 1.408% with polysulfide-based electrolyte and Cu_2S counter electrode. The proper adjustment of the absorption to the solar spectrum was realized by changes in the composition and size of QDs (Elibol, 2020). Quaternary alloy quantum dots of Cu-In-Sn-Se can also present a wide range of absorbed light wavelength. Due to the introduction of quaternary alloy dots to the photocells, the reduction in charge recombination and improvement in the collection of electrons was achieved. The PCE of the photocells reached 5.64% and 6.7% with additional passivation of photoanode surface by ZnS (Liu et al., 2021). The usefulness of passivation of photoanode by ZnS or ZnSe was also demonstrated experimentally earlier, in QDSSC co-sensitized with CdS and CdSe quantum dots. The incorporation of passivating layer postpones the charge recombination by extracting the hole from QD and moving it to the valence band of the passivation layer, which has a lower potential (Huang et al., 2016). In consequence, a strong suppression of recombination and an enhancement in cell efficiency are observed.

5.3.3 Core-shell structure

Another strategy is the usage of quantum dots with core-shell structure composed of different materials; the core-shell structure can be divided into two types: in type I core-shell structure, the combination of a low-bandgap core and a wide-bandgap shell prevents the recombination of charge carriers. In type II, the shell conduction band or shell valence band partially overlaps the bandgap of the core, which favors a longer carrier lifetime. Figure 5.5 shows the structure and energy band alignment of type I and type II core-shell quantum dots that are constituted with the properly matched materials. The presented alignment of the energy levels facilitates charge separation and electron injection from QD to the wide-bandgap semiconductor layer.

The application of type I CdSeTe/CdS quantum dots, with a wide-bandgap shell protecting the core, limited the density of surface trap defects and a delivered maximum PCE of 9.48% (Yang et al., 2015). The CdS shell with an optimized thickness contributed to the enhancement of PCE by 13% compared to QDSSC with plain CdSeTe QDs.

Another example of type I core-shell structure QDs is InAs/ZnSe quantum dots, in which the shell stabilizes the core and prevents surface oxidation. The InAs/ZnSe QDs demonstrated a large absorption coefficient and multiple exciton generation, which led to the 2.7% efficiency of the photocells (Lee et al., 2015).

Owing to a broad range of absorption, fast charge separation and suppressed recombination, type II CdTe/CdSe core-shell QDs (CdTe with $E_g = 1.5\,eV$ and CdSe with $E_g = 1.74\,eV$) exhibited a PCE of 6.76% (Wang et al., 2013). The modification of such a core-shell structure is ZnTe/CdSe QD (the ZnTe core exhibits E_g of 2.26 eV), which proved excellent optoelectronic properties. Due to the large conduction band offset, which resulted in an increase in the photovoltage, the maximum PCE obtained was 7.17% (Jiao et al., 2015). In another experimental research, the CdSe/Cu$_2$Se core-shell quantum dots showed a suitable crystal structure of nanoparticles and absorption of up to 700 nm. The application of CdSe/Cu$_2$Se QDs as a sensitizer on TiO$_2$ with sulfide/polysulfide electrolyte and Cu$_2$S counter electrode led to a PCE of 4.83%. The photoanode in the photocells was coated by ZnS to suppress recombination and improve electron transport (Simi et al., 2021).

The CdTe-free core-shell Cu$_2$GeS$_3$/InP type II QDs provided the optimization of bands alignment, which resulted in the fast electron injection rate, efficient charge

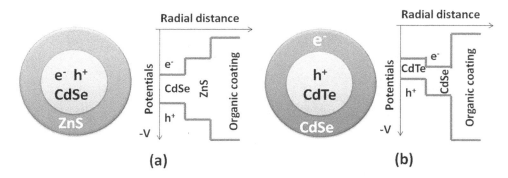

Figure 5.5 Structure and energy bands alignment of (a) type I and (b) type II core-shell quantum dots. (Copied under Creative Commons Attribution 3.0 License from Jasim (2015).)

separation, limitation of carriers recombination and, in consequence, a five times better efficiency than the pristine Cu_2GeS_3 (Jamshidi Zavaraki et al., 2018).

An approach with metal/QD core-shell structure, realized in Mg/CdS core-shell QDs with ZnS passivation on TiO_2 photoelectrode, demonstrated an enhancement in charge separation. Due to the effective light harvesting and electron-hole separation, a PCE of 7.47% was obtained (Khodam et al., 2019).

5.3.4 Co-sensitization

The beneficial strategy that is aimed at the proper match of the QDSSC absorption range to the solar spectrum is co-sensitization by various types of nanoparticles, the selection of which should take into account their bandgap values that differ from the bulk material bandgaps. The research on the combination of CdS, CdTe, CdSe and PbS quantum dots as sensitizers in QDSSC proved the enhancement in photon absorption and in consequence a greater number of generated excitons (Huang et al., 2016; Chung et al., 2021).

The promising results of co-sensitization were demonstrated in the photocells employing the blend of CdS and InSb quantum dots, which improved the light absorption and limited the recombination losses. The enhancements of current density to 18.58 mA/cm^2, open-circuit voltage to 0.533 V and efficiency to 4.94% were observed in comparison with CdS QDs-based photocells in which a current density of 16.98 mA/cm^2, open-circuit voltage 0.454 V and efficiency of up to 3.52% were obtained (Archana et al., 2020).

The co-sensitization effect can also be achieved by the utilization of carbon nanomaterials such as graphene, graphene quantum dots (GQD), graphene oxide (GO) or reduced graphene oxide (RGO) together with semiconductor nanoparticles. The example of such an approach is the employment of the CdSe QDs mixed with GO as a photoactive layer, which enhanced the sensitization effect as well as further charge

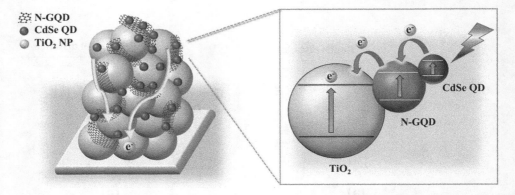

Figure 5.6 Nitrogen-doped graphene quantum dots (N-GQD) incorporated in the layer of TiO_2 nanoparticles (TiO_2 NP) sensitized with CdSe quantum dots (CdSe QD). The inset shows the relative alignment of energy levels. (Reprinted from Jo et al. (2021). Copyright with permission from Elsevier.)

separation and transport and resulted in the improvement in the photocurrent by 150% (Lightcap & Kamat, 2012).

The co-sensitization with carbon nanomaterial was realized in the QDSSC based on TiO_2 photoanode covered with nitrogen-doped graphene quantum dots (N-GQD) synthetized using the hydrothermal method and CdSe nanoparticles grown by the SILAR technique (Jo et al., 2021). Figure 5.6 schematically depicts the structure and energy bands in the CdSe-sensitized TiO_2 film with introduced graphene quantum dots. In addition to the role of co-sensitizer, N-GQD acted also as a passivation layer which suppresses charge recombination, improves charge transport and light absorption, as well as offers nucleation sites for semiconductor quantum dots. The QDSSC co-sensitized with N-GQD achieved a PCE of 4.8% compared to 3.92% for devices sensitized only with CdSe.

5.4 MULTIPLE EXCITON GENERATION

The theoretical limit of solar cell efficiency determined by Shockley and Queisser is based on the assumption that the absorption of one photon of energy greater than the bandgap results in the generation of one electron-hole pair and the excess of photon energy is lost as heat (Shockley & Queisser, 1961). According to this theory, the photogenerated electrons and holes are cooled from initial hot state by emission of phonons and quickly relax to the band edges. This process can be followed by a much slower radiative or non-radiative recombination.

A phenomenon that can help to overcome the theoretical efficiency limit of a single-junction solar cell is the multiple exciton generation (MEG), which was observed experimentally by the transient absorption spectroscopy method in colloidal quantum dots. In the process of multiple exciton generation, the absorption of a photon with energy at least twice as high as the bandgap can result in the generation of more than one electron-hole pair. This groundbreaking observation, showing that the dynamics of the relaxation of hot carriers in semiconductor nanostructures is influenced by quantum effects, was successfully utilized in QDSSC (Nozik, 2002). The principle of MEG in a quantum dot is shown in Figure 5.7. Owing to this phenomenon, the introduction of semiconductor nanoparticles into solar cells enables us to significantly reduce the loss of excess photon energy. As a result, the external quantum efficiency (EQE – the ratio of collected electrons to the number of incident photons at a given light wavelength) and internal quantum efficiency (IQE, which is EQE corrected for reflection and absorption losses) of the photocell can exceed 100%.

The MEG occurring in quantum dot sensitizers was shown to provide more efficient harvesting of the excess energy than impact ionization, which is the similar process observed in bulk semiconductors. The experimental research demonstrated that the absorption of photon of energy four times greater than the bandgap of PbSe QD resulted in the generation of three excitons, which means 300% quantum yield (Ellingson et al., 2005). Further progress in the charge carrier generation was also achieved in lead selenide (PbSe) nanocrystals with 0.636 eV bandgap, in which a photon with an energy of 7.8 energy gaps generated seven excitons; thus, the EQE reached a spectacular value of 700% and the loss of energy was only around 10% (Schaller et al., 2006).

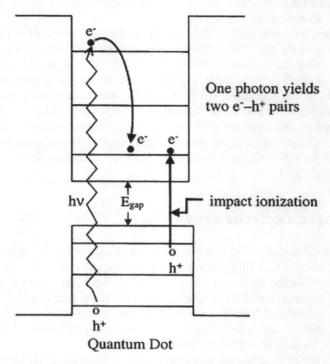

Figure 5.7 Scheme of multiple electron-hole pair creation by one photon. The impact ionization indicated in the figure is called multiple exciton generation. (Reprinted from Nozik (2002). Copyright with permission from Elsevier.)

The phenomenon of multiple exciton generation was also observed in the QDSSC with the following configuration: ITO/ZnO/PbSe QDs/Au, for which an EQE exceeding 100% was obtained under zero external bias. The solar cells sensitized with PbSe quantum dots provided a photocurrent improvement arising from MEG associated with a maximum external quantum efficiency of 114% and an internal quantum efficiency of 130% (Semonin et al., 2011).

5.5 SENSITIZATION PROCESS – DEPOSITION OF QUANTUM DOTS

The quantum dots that sensitize the wide-bandgap semiconductor nanostructure in QDSSC can be deposited in situ, when the growth of quantum dots undergoes directly on mesoporous layer, or ex situ, when the ready, synthetized quantum dots are adsorbed on the surface of the mesoporous layer.

5.5.1 In situ deposition

An easy and non-expensive in situ method is the successive ionic layer adsorption and reaction (SILAR), which enables to control the load of nanoparticles by changing the number of cycles, immersion time and concentration of QD precursor solution. The SILAR method was successfully applied in order to sensitize TiO_2 with widely used CdS or PbS quantum dots (Raphael et al., 2017). To realize sensitization with PbS QDs, the TiO_2 photoanode was dipped in the Pb^{+2} precursor ($Pb(NO_3)_2$ solution in methanol), and after rinsing with methanol, it was immersed in the S^{2-} precursor (Na_2S solution in methanol and deionized water) (Basit et al., 2020). The described cycle has to be repeated three times to obtain the optimum amount of sensitizing quantum dots. Figure 5.8 shows the high-quality TEM images of the PbS quantum dots obtained by means of the SILAR method on TiO_2 nanoparticles. A similar procedure with ethanol solution of $Cd(NO_3)_2$ used at the first step of the SILAR method provides CdS quantum dots for application in QDSSC (Zhang et al., 2018). The in situ co-sensitization of TiO_2, e.g., with CdS-ZnS QDs, was also realized by the SILAR technique. In the first stage, the prefabricated TiO_2 photoelectrode was immersed in the $Cd(NO_3)_2$ solution in ethanol and then in the Na_2S aqueous solution to obtain CdS QDs, and in the second stage, the procedure was repeated with $Zn(OAc)_2$ in ethanol and the Na_2S aqueous solution, which resulted in the growth of ZnS QDs. Finally, multiple layers of CdS-ZnS QDs were deposited on specially prepared glass covered by FTO and dandelion-like TiO_2 microspheres and were investigated by XRD, SEM, EDAX and HRTEM analyses (Aslam et al., 2015).

Another in situ method is the chemical bath deposition, the advantages of which include simple system, rate optimization and sufficient uniform coverage of the substrate. Using this technique, the Mn^{2+} doped CdSe quantum dots were deposited on TiO_2 electrodes by immersion in the solution containing $Mn(CH_3COO)_2$ as a source of Mn^{2+} ions (Zhang et al., 2018). The SEM, TEM and EDS characterization methods enabled the verification of the morphology and composition of the obtained quantum dots.

5.5.2 Ex situ deposition

The ex situ methods of sensitization involve the use of previously prepared quantum dots dispersion. First, quantum dots of desired size and structure have to be synthesized; then, they can be deposited on TiO_2 by spin coating or simple immersion of the electrode in a colloidal solution of quantum dots. Such an approach allows for sensitization with QDs of widely variable composition. The example includes the InAs/Se core-shell QDs, in which the InAs core was synthesized in a precursors mixture of $In(OAc)_3$ and $(TMS)_3As$ with myristic acid as surfactant and then the shell was grown by the slow addition of a solution containing Se. After the deposition of two-monolayer shell, the QDs of 5 nm size were obtained and their properties were tested by TEM, XRD and spectrophotometer. The prepared colloidal solution of QDs was spin-coated on the TiO_2 photoanode (Lee et al., 2015). A similar procedure was employed to fabricate the CdSe-Cu_2Se core-shell structure. Stirring of the precursors under argon atmosphere and controlled temperature was followed by purification, centrifugation of QDs and dispersion in the solvent. The deposition of QDs on TiO_2 was realized ex situ by dipping the electrode in the colloidal solution of nanoparticles (Simi et al., 2021). The metal-doped QDs were also obtained by using ex situ methods. $CdCl_2$ and Na_2S mixed with PVA solution resulted in the growth of CdS quantum dots, and manganese acetate hydrate was added in order to dope the QDs with Mn. Glass/FTO/TiO_2

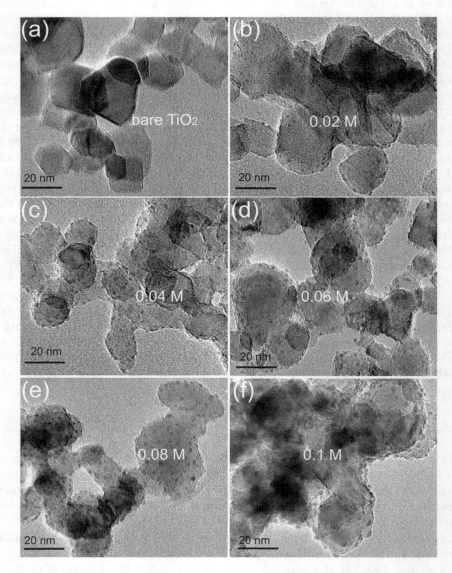

Figure 5.8 Transmission electron microscope images showing (a) bare TiO_2 nanoparticles and (b–f) TiO_2 nanoparticles with PbS quantum dots grown by the SILAR method. The load of PbS depends on the concentration of QDs precursor solution, which ranges from 0.02 to 0.1 M. (Copied under Creative Commons Attribution 4.0 International License from Tian et al. (2016).)

electrode was dip-coated by immersion in solution containing the QDs (Ganguly & Nath, 2020). The promising modification of this ex situ method includes the incorporation of bifunctional linkers between TiO_2 surface and sensitizing quantum dots. Different tested linkers, e.g., phenyl-MBA, thiophene-TPA and alkyl-TGA, were deposited on TiO_2 by immersion in a linker solution prior to the sensitization with QDs. The functionalization of TiO_2 electrodes was followed by dipping in dispersion of CdSe QDs (Zhang et al., 2021).

The alternative to direct adsorption of QDs from solution is the electrophoresis method, in which prefabricated TiO_2-coated FTO glass serves as the positive electrode and the glass/FTO as the negative electrode connected to the DC power supply. Both electrodes are immersed in a solution of QDs, and during the electrodeposition process, the electrodes are covered by the dots. The described electrophoretic deposition can be used to sensitize the mesoporous TiO_2 layer by colloidal quantum dots, e.g., CdSeTe or $CuInS_2$ alloy QDs (Raphael et al., 2017; Purcell-Milton et al., 2019; Elibol, 2020).

5.6 OTHER COMPONENTS OF QDSSC

5.6.1 Electron transport layer

One of the key parts of QDSSC is the electron transport layer (ETL) on the anode whose conduction band should be positioned lower in energy than the conduction band of light-absorbing quantum dots, in order to effectively accept photogenerated electrons from QDs. Other requirements for ETL are photostability and sufficient conductivity, which is a crucial parameter for efficient transport of electrons and their collection by an adequate electrode. Like in dye-sensitized solar cells, TiO_2 in the form of mesoporous layer of sintered nanoparticles is the most frequently used electron transport material in QDSSC. However, studies show that the utilization of one-dimensional nanomaterials, such as nanorods, nanowires or nanotubes, is favorable, since they provide direct channels for charge carriers and thus exhibit better charge transport properties and lower recombination rates (Aslam et al., 2015). Other features, which can also be included among their advantages, are the large surface area for the adsorption of the sensitizer and facile preparation methods. The example is the TiO_2 nanowire array prepared by the hydrothermal method treated with $TiCl_4$, which reduces the recombination and resistance as well as provides excellent electron transfer properties and electron collection capability. The solar cells based on the TiO_2 nanowires pre-treated with $TiCl_4$ and sensitized with CdS_xSe_{1-x} quantum dots delivered $J_{SC} = 18.93\,mA/cm^2$, $V_{OC} = 0.78\,V$, $FF = 51.1\%$ and efficiency of 7.54% (Li et al., 2019).

A promising solution is also the introduction of graphene into ETL. The nanocomposite consisting of TiO_2 rutile nanorods prepared by means of the hydrothermal method and coated with graphene oxide, mixed with TiO_2 anatase nanoparticles, was used in QDSSC, which provided a PCE of 2.2% (Huang et al., 2019).

A wide-bandgap n-type semiconductor substitute for TiO_2 in the role of ETL can be nanostructural ZnO, which presents a proper conductivity and efficiently adsorbs quantum dots. The experimental research showed that the layer of randomly oriented ZnO nanowires facilitates charge transfer in QDSSC. The devices with the ZnO nanowires obtained by electrochemical deposition sensitized by SbS_3 QDs attained an efficiency of 2.43% (Li et al., 2020). To limit the charge carrier recombination, a hierarchical TiO_2/ZnO structure was prepared, consisting of porous TiO_2 that can adsorb quantum dots and one-dimensional ZnO nanowires that can scatter light and provide elongated pathways for electron transport. The PCE exceeding 3% was obtained by a photocell consisting of CdSe QDs-sensitized photoanode, Pt counter electrode and polysulfide electrolyte (Du et al., 2019).

Another approach to the improvement of ETL is the development of composites containing carbon nanostructures. An example of such a composite is TiO_2 doped with neodymium and mixed with one-dimensional graphene nanoribbons. The Nd doping was realized by mixing of precursors followed by annealing and then the application of the solvothermal method to synthesize the nanocomposite. The study demonstrated that the incorporation of the dopant implies a longer charge carrier lifetime and bandgap narrowing, whereas graphene nanoribbons favor fast electron extraction and improved charge separation. The described modification of photoanode sensitized with CdS QDs resulted in a PCE of 5.8%, which is more than twice as much as for bare TiO_2 (Akash et al., 2021).

The TiO_2 ETL, which is highly porous in nature, does not tightly cover the surface of the FTO film; thus, it is advisable to introduce an interface layer to prevent the contact between the bare FTO surface and the electrolyte in order to suppress the charge recombination at the FTO/electrolyte interface. Reduced graphene oxide (RGO), which can be easily deposited by spraying of RGO dispersion in ethanol directly on FTO, was found to be useful as the layer that blocks the back transport of electrons (Wang et al., 2019). The introduction of a thin RGO layer under ETL coverage contributed to the improvement of photovoltaic performance of QDSSC based on CdS nanoparticles and led to a PCE of 4.7%. This novel approach differs from other studies, in which RGO was an additive in porous TiO_2 layer or the counter electrode.

5.6.2 Electrolyte

The electrolyte in QDSSC withdraws holes from quantum dots, after exciton separation followed by electron transfer to the ETL, by donating the electrons to QD. Polysulfide liquid electrolyte with S^{2-}/Sn^{2-} redox couple, which replaced traditional iodide-based electrolyte in order to prevent the corrosion of quantum dots, is widely used and efficient in QDSSC (Li et al., 2019; Du et al., 2019; Basit et al., 2020). The modification of polysulfide electrolyte by the addition of tetraethyl orthosilicate proved effective performance with CdSe and CdSeTe QDs, and an impressive PCE of up to 12.34% was achieved with the use of quaternary Zn-Cu-In-Se QDs (Yu et al., 2017).

Volatilization and leakage issues usually accompany the use of liquid electrolytes; therefore, a development of gel or solid-state electrolytes could be beneficial. Gel polymeric electrolytes are cheap and stable, and they present conductivity similar to liquid electrolytes and sufficient penetration into the mesoporous photoanode. The introduction of agar-based gel polymeric electrolyte led to the improvement of the QDSSC stability, IPCE of over 51% and PCE close to 3% with sensitization by CdS or $CuInS_2$ QDs (Raphael et al., 2017). Outstanding stability, retaining 67% of initial efficiency after 504 h, was achieved by the QDSSC with solid-state electrolyte based on benzimidazolium salt. Upon sensitization with CdS/CdSe quantum dots, the photocells delivered a PCE of 4.26%, $J_{SC}=12.58\,mA/cm^2$, $V_{OC}=0.6\,V$ and $FF=56.44\%$ (Dang et al., 2017). The utilization of oxide nanoceramics as a highly porous solid-state electrolyte also proved excellent stability of up to 60 days under ambient conditions as well as improvement in V_{OC} and reduction in series resistance in comparison with polysulfide-based electrolytes. The doped ceramics used in CdS-sensitized photocell facilitated a quick charge transport, suppressed the recombination and led to a PCE of 2.76% (Kusuma & Balakrishna, 2020).

5.6.3 Counter electrode

A counter electrode, which collects electrons from an external circuit and reduces the electrolyte, is also an important part of QDSSC. Commonly used counter electrodes coated with Pt or metal sulfides such as CoS, PbS and Cu_2S presented satisfactory performance in contact with polysulfide liquid electrolyte; however, the improvement of conductivity and electrocatalytic activity is still necessary (Li et al., 2019; Basit et al., 2020; Simi et al., 2021).

The search for materials of better conductivity led to the employment of composites based on carbon nanostructures, which are known for providing superior electronic and mechanical properties. The counter electrode was prepared by covering the glass/FTO with carbon allotropes (graphene, graphene oxide and carbon nanotubes) by using the electrophoresis method and deposition of CoS nanoparticles doped with Sr, Ba, Mg and Ca. The photocells consisting of a TiO_2 photoanode co-sensitized with CdS/CdSe/ZnS nanoparticles, a polysulfide electrolyte and the nanocomposite-coated counter electrode provided the highest efficiencies of 2.81% for Sr-CoS/GO and 2.26% for Ba-CoS/GO, which indicates the improvement compared to bare CoS (Khalili et al., 2017). The boost of PCE up to 8.28% was achieved by the QDSSC with counter electrode covered by integrated architecture CNT/RGO/MoCuSe. In this case, the application of the simple hydrothermal method enabled us to fabricate stable and efficient coverage of counter electrode (Gopi et al., 2018). Reduced graphene oxide was also incorporated in a counter electrode of QDSSC with Cu_2S gel polymer electrolyte, delivering a PCE of 2.97% (Raphael et al., 2017). The same mentioned facile hydrothermal synthesis was used to grow a nanocomposite consisting of CoS nanoparticles and $Ti_3C_2T_x$ nanosheets, which demonstrated high electrochemical performance and stability. The $J_{SC}=21.29\,mA/cm^2$, $V_{OC}=0.671$ V and FF = 56.6% were delivered, and a PCE exceeding 8% was attained owing to high porosity, numerous catalytic active sites and fast electron transfer of the developed counter electrode material (Chen et al., 2021).

5.7 SUMMARY

In quantum dot solar cells, mainly CdS, CdSe, CdTe and PbS quantum dots are used as sensitizers of TiO_2 layer. Numerous exceptional properties of quantum dots including tunable absorption range result from their quantum confinement. Among these features, a special place is occupied by the ability of multiple exciton generation, which may improve the efficiency and potentially allows overcoming the theoretical efficiency limit. As a result of multiple exciton generation, the external quantum efficiency of QDSSC reached 700%; however, the impact on the final photovoltaic performance is not as significant due to the short lifetime of the photogenerated excitons. In fact, the current record power conversion efficiency of quantum dot-sensitized solar cells is below 13% and further efforts should be focused on the suppression of deleterious recombination, as well as the elimination of Cd and Pb as components of the dots for environmental safety reasons.

REFERENCES

Akash, S., Shwetharani, R., Kusuma, J., Akhil, S. & Balakrishna, R.G. (2021) Highly efficient and durable electron transport layer for QDSSC: an integrated approach to address recombination losses. *Journal of Alloys and Compounds* 162740. doi: 10.1016/j.jallcom.2021.162740.

Archana, T., Vijayakumar, K., Subashini, G., Grace, A.N., Arivanandhan, M. & Jayavel, R. (2020) Effect of co-sensitization of InSb quantum dots on enhancing the photoconversion efficiency of CdS based quantum dot sensitized solar cells. *Royal Society of Chemistry Advances* 10, 14837–14845. doi: 10.1039/c9ra10118g.

Arivarasan, A., Bharathi, S., Arasi, S.E., Arunpandiyan, S., Revathy, M.S. & Jayavel, R. (2020) Investigations of rare earth doped CdTe QDs as sensitizers for quantum dots sensitized solar cells. *Journal of Luminescence* 219, 116881. doi: 10.1016/j.jlumin.2019.116881.

Aslam, M.M., Ali, S.M., Fatehmulla, A., Farooq, W.A., Atif, M., Al-Dhafiri, A.M. & Shar, M.A. (2015) Growth and characterization of layer by layer CdS–ZnS QDs on dandelion like TiO_2 microspheres for QDSSC application. *Materials Science in Semiconductor Processing* 36, 57–64. doi: 10.1016/j.mssp.2015.03.030.

Basit, M.A., Abbas, M.A., Naeem, H.M., Ali, I., Jang, E., Bang, J.H. & Park, T.J. (2020) Ultrathin TiO_2-coated SiO_2 nanoparticles as light scattering centers for quantum dot-sensitized solar cells. *Materials Research Bulletin* 127, 110858. doi: 10.1016/j.materresbull.2020.110858.

Chen, X., Zhuang, Y., Shen, Q., Cao, X., Yang, W. & Yang, P. (2021) In situ synthesis of $Ti_3C_2T_x$ MXene/CoS nanocomposite as high performance counter electrode materials for quantum dot-sensitized solar cells. *Solar Energy* 226, 236–244. doi: 10.1016/j.solener.2021.08.053.

Chung, N., Nguyen, P.T., Tung, H.T. & Phuc, D.H. (2021) Quantum dot sensitized solar cell: photoanodes, counter electrodes, and electrolytes. *Molecules* 26(9), 2638. doi: 10.3390/molecules26092638.

Dang, R., Wang, Y., Zeng, J., Huang, Z., Fei, Z. & Dyson, P.J. (2017) Benzimidazolium salt-based solid-state electrolytes afford efficient quantum-dot sensitized solar cells. *Journal of Materials Chemistry A* 5, 13526–13534. doi: 10.1039/C7TA02925J.

Du, X., Li, W., Zhao, L., He, X., Chen, H. & Fang, W. (2019) Electron transport improvement in CdSe-quantum dot solar cells using ZnO nanowires in nanoporous TiO_2 formed by foam template. *Journal of Photochemistry and Photobiology A: Chemistry* 371, 144–150. doi: 10.1016/j.jphotochem.2018.10.054.

Elibol, E. (2020) Effects of different counter electrodes on performance of CdSeTe alloy QDSSC. *Solar Energy* 197, 519–526. doi: 10.1016/j.solener.2020.01.035.

Ellingson, R.J., Beard, M.C., Johnson, J.C., Yu, P., Micic, O.I., Nozik, A.J., Shabaev, A. & Efros, A.L. (2005) Highly efficient multiple exciton generation in colloidal PbSe and PbS quantum dots. *Nano Letters* 5(5), 865–871. doi: 10.1021/nl0502672.

Ganguly, A. & Nath, S.S. (2020) Mn-doped CdS quantum dots as sensitizers in solar cells. *Materials Science and Engineering: B* 255, 114532. doi: 10.1016/j.mseb.2020.114532.

Gopi, C.V.V.M., Singh, S., Reddy, A.E. & Kim, H.-J. (2018) CNT@rGO@MoCuSe composite as an efficient counter electrode for quantum dot-sensitized solar cells. *ACS Applied Materials and Interfaces* 10(12), 10036–10042. doi: 10.1021/acsami.7b18526.

Huang, F., Zhang, Q., Xu, B., Hou, J., Wang, Y., Massé, R.C., Peng, S., Liu, J. & Cao, G. (2016) A comparison of ZnS and ZnSe passivation layers on CdS/CdSe co-sensitized quantum dot solar cells. *Journal of Materials Chemistry A* 4, 14773–14780. doi: 10.1039/C6TA01590E.

Huang, T., Zhang, X., Wang, H., Chen, X., Wen, L., Huang, M., Zhong, Y., Luo, H., Tang, G. & Zhou, L. (2019) Improved CdS QDSSCs with graphene and anatase-rutile TiO_2 composite as photoanodes. *Superlattices and Microstructures* 126, 17–24. doi: 10.1016/j.spmi.2018.12.007.

Jamshidi Zavaraki, A., Huang, J., Ji, Y. & Ågren, H. (2018) Low toxic Cu_2GeS_3/InP quantum dot sensitized infrared solar cells. *Journal of Renewable and Sustainable Energy* 10, 043710. doi: 10.1063/1.5044608.

Jasim, K.E. (2015) Quantum dots solar cells, solar cells: new approaches and reviews. In: Kosyachenko, L.A. (eds) *Solar Cells*. IntechOpen: London. https://www.intechopen.com/chapters/47671.

Jiao, S., Shen, Q., Mora-Seró, I., Wang, J., Pan, Z., Zhao, K., Kuga, Y., Zhong, X. & Bisquert, J. (2015) Band engineering in core/shell ZnTe/CdSe for photovoltage and efficiency enhancement in exciplex quantum dot sensitized solar cells. *ACS Nano* 9(1), 908–915. doi: 10.1021/nn506638n.

Jo, I.-R., Lee, Y.-H., Kim, H. & Ahn, K.-S. (2021) Multifunctional nitrogen-doped graphene quantum dots incorporated into mesoporous TiO_2 films for quantum dot-sensitized solar cells. *Journal of Alloys and Compounds* 870, 159527. doi: 10.1016/j.jallcom.2021.159527.

Khalili, S.S, Dehghani, H. & Afrooz, M. (2017) Composite films of metal doped CoS/carbon allotropes; efficient electrocatalyst counter electrodes for high performance quantum dot-sensitized solar cells. *Journal of Colloid and Interface Science* 493, 32–41. doi: 10.1016/j.jcis.2017.01.005.

Khodam, F., Amani-Ghadim, A.R. & Aber, S. (2019) Mg nanoparticles core-CdS QDs shell heterostructures with ZnS passivation layer for efficient quantum dot sensitized solar cell. *Electrochimica Acta* 308, 25–34. doi: 10.1016/j.electacta.2019.03.228.

Kouhnavard, M., Ikeda, S., Ludin, N.A., Ahmad Khairudin, N.B., Ghaffari, B.V., Mat-Teridi, M.A., Ibrahim, M.A., Sepeai, S. & Sopian, K. (2014) A review of semiconductor materials as sensitizers for quantum dot-sensitized solar cells. *Renewable and Sustainable Energy Reviews* 37, 397–407. doi: 10.1016/j.rser.2014.05.023.

Kusuma, J. & Balakrishna, R.G. (2020) Ceramic grains: highly promising hole transport material for solid state QDSSC. *Solar Energy Materials and Solar Cells* 209, 110445. doi: 10.1016/j.solmat.2020.110445.

Lee, S.H., Jung, C., Jun, Y. & Kim, S.-W. (2015) Synthesis of colloidal InAs/ZnSe quantum dots and their quantum dot sensitized solar cell (QDSSC) application. *Optical Materials* 49, 230–234. doi: 10.1016/j.optmat.2015.09.027.

Li, Z., Wang, Y.F.W., Wang, X.W., Yang, Z. & Zeng, J.H. (2018) Doping as an effective recombination suppressing strategy for performance enhanced quantum dots sensitized solar cells. *Materials Letters* 221, 42–45. doi: 10.1016/j.matlet.2018.03.058.

Li, D., Jiang, Z., Xia, Q. & Yao, Z. (2019) Pre- or post-$TiCl_4$ treated TiO_2 nano-array photoanode for QDSSC: Ti^{3+} self-doping, flat-band level and electron diffusion length. *Applied Surface Science* 491, 319–327. doi: 10.1016/j.apsusc.2019.06.166.

Li, W., Peng, Z., Sun, Z., Liu, Z., Chen, J., Qiu, W., Chen, J. & Zhang, S. (2020) Orientation modulation of ZnO nanorods on charge transfer performance enhancement for Sb_2S_3 quantum dot sensitized solar cells. *Journal of Alloys and Compounds* 816, 152628. doi: 10.1016/j.jallcom.2019.152628.

Lightcap, I.V. & Kamat, P.V. (2012) Fortification of CdSe quantum dots with graphene oxide: excited state interactions and light energy conversion. *Journal of American Chemical Society* 134(16), 7109–7116. doi: 10.1021/ja3012929.

Liu, S., Fan, R., Zhao, Y., Yu, M., Fu, Y., Li, L., Li, Q., Liang, B. & Zhang, W. (2021) Assembly of Cu–In–Sn–Se quantum dot-sensitized TiO_2 films for efficient quantum dot–sensitized solar cell application. *Materials Today Energy* 21, 100798. doi: 10.1016/j.mtener.2021.100798.

Nozik, A.J. (2002) Quantum dot solar cells. *Physica E: Low-Dimensional Systems and Nanostructures* 14(1–2), 115–120. doi: 10.1016/S1386-9477(02)00374-0.

Purcell-Milton, F., Curutchet, A. & Gun'ko, Y. (2019) Electrophoretic deposition of quantum dots and characterisation of composites. *Materials* 12(24), 4089. doi: 10.3390/ma12244089.

Raphael, E., Jara, D.H. & Schiavon, M.A. (2017) Optimizing photovoltaic performance in $CuInS_2$ and CdS quantum dot-sensitized solar cells by using an agar-based gel polymer electrolyte. *Royal Society of Chemistry Advances* 7(11), 6492–6500. doi: 10.1039/C6RA27635K.

Sahu, A., Garg, A. & Dixit, A. (2020) A review on quantum dot sensitized solar cells: past, present and future towards carrier multiplication with a possibility for higher efficiency. *Solar Energy* 203, 210–239. doi: 10.1016/j.solener.2020.04.044.

Schaller, R.D., Sykora, M., Pietryga, J.M. & Klimov, V.I. (2006) Seven excitons at a cost of one: redefining the limits for conversion efficiency of photons into charge carriers. *Nano Letters* 6, (3), 424–429. doi: 10.1021/nl052276g.

Semonin, O.E., Luther, J.M., Choi, S., Chen, H.-Y., Gao, J., Nozik, A.J. & Beard, M.C. (2011) Peak external photocurrent quantum efficiency exceeding 100% via MEG in a quantum dot solar cell. *Science* 334(6062), 1530–1533. doi: 10.1126/science.1209845.

Shockley, W. & Queisser, H.J. (1961) Detailed balance limit of efficiency of p-n junction solar cells. *Journal of Applied Physics* 32, 510. doi: 10.1063/1.1736034.

Simi, N.J., Bernadsha, B., Thomas, A. & Ison, V.V. (2021) Quantum dot sensitized solar cells using type-II $CdSe-Cu_2Se$ core-shell QDs. *Results in Optics* 4, 100088. doi: 10.1016/j.rio.2021.100088.

Song, H., Lin, Y., Zhang, Z., Rao, H., Wang, W., Fang, Y., Pan, Z. & Zhong, X. (2021) Improving the efficiency of quantum dot sensitized solar cells beyond 15% via secondary deposition. *Journal of American Chemical Society* 143(12), 4790–4800. doi: 10.1021/jacs.1c01214.

Tian, J., Shen, T., Liu, X., Fei, C., Lv, L. & Cao, G. (2016) Enhanced performance of PbS quantum-dot-sensitized solar cells via optimizing precursor solution and electrolytes. *Scientific Reports* 6, 23094. doi: 10.1038/srep23094.

Tianxing, L., Zou, X. & Zhou, H. (2014) Effect of Mn doping on properties of CdS quantum dot-sensitized solar cells. *International Journal of Photoenergy* 569763. doi: 10.1155/2014/569763.

Wang, J., Mora-Seró, I., Pan, Z., Zhao, K., Zhang, H., Feng, Y., Yang, G., Zhong, X. & Bisquert, J. (2013) Core/shell colloidal quantum dot exciplex states for the development of highly efficient quantum-dot-sensitized solar cells. *Journal of American Chemical Society* 135(42), 15913–15922. doi: 10.1021/ja4079804.

Wang, L., Feng, J., Tong, Y. & Liang, J. (2019) A reduced graphene oxide interface layer for improved power conversion efficiency of aqueous quantum dots sensitized solar cells. *International Journal of Hydrogen Energy* 44(1), 128–135. doi: 10.1016/j.ijhydene.2018.01.155.

Yang, J., Wang, J., Zhao, K., Izuishi, T., Li, Y., Shen, Q. & Zhong, X. (2015) CdSeTe/CdS type-I core/shell quantum dot sensitized solar cells with efficiency over 9%. *Journal of Physical Chemistry C* 119(52), 28800–28808. doi: 10.1021/acs.jpcc.5b10546.

Yu, J., Wang, W., Pan, Z., Du, J., Ren, Z., Xue, W. & Zhong, X. (2017) Quantum dot sensitized solar cells with efficiency over 12% based on tetraethyl orthosilicate additive in polysulfide electrolyte. *Journal of Materials Chemistry A* 5(27), 14124–14133. doi: 10.1039/C7TA04344A.

Zhang, C., Liu, S., Liu, X., Deng, F., Xiong, Y. & Tsai, F.C. (2018) Incorporation of Mn^{2+} into CdSe quantum dots by chemical bath co-deposition method for photovoltaic enhancement of quantum-dot-sensitized solar cells. *Royal Society Open Science* 5(3), 171712. doi: 10.1098/rsos.171712.

Zhang, D., Zhang, S., Fang, Y., Xie, D., Zhou, X. & Lin, Y. (2021) Effect of linkers with different chemical structures on photovoltaic performance of CdSe quantum dot-sensitized solar cells. *Electrochimica Acta* 367, 137452. doi: 10.1016/j.electacta.2020.137452.

Chapter 6

Environmental Impact of Emerging Photovoltaics

6.1 INTRODUCTION

Emerging photovoltaics gathers much attention due to the ability to achieve beneficial cost-to-efficiency ratio and potential for a wide variety of applications. Like all new technologies that will be implemented in the future and brought to the market, the photovoltaics of third generation requires environmental impact analysis.

The photovoltaic cells belonging to the third generation are characterized by a wide range of options for the selection of various alternative materials employed as a photoactive layer, electron and hole transport layers, as well as electrodes, which is an important feature from the environmental point of view. Thus, with sustainable development in mind, the emerging photovoltaic technologies should be based on conscious selection of materials and processes, which are economically and environmentally friendly. In order to determine the environmental burden of particular types of solar cells belonging to the class of emerging photovoltaics and provide the comparison between different cells in this regard, it is necessary to apply an adequate method enabling us to perform multifaceted, long-term analysis. The method commonly applied to evaluate photovoltaics in the context of sustainability is the life cycle assessment, which provides a complex picture of the environmental performance from the production stage of the photocells to their disposal. According to assumptions, the entire analysis starts from raw materials acquisition, and includes degradation during lifetime, transportation and end-of-life scenario. However, if the study is devoted to a technology in an early phase of development, like emerging photovoltaics, some difficulties are involved, such as lack of real data of long-term performance or disposal options.

In this chapter, the application of the life cycle assessment method to organic, dye-sensitized, perovskite and quantum dot-sensitized solar cells is presented. The described case studies include small single-laboratory photocells as well as minimodules mounted in different places. A comparison of the third generation photovoltaics to the mature silicon technology in the context of environmental impact is also made.

6.2 LIFE CYCLE ASSESSMENT OF PHOTOVOLTAIC TECHNOLOGIES

Life cycle assessment (LCA) is a comprehensive method that enables quantitative evaluation of the environmental impact of a product or process over the entire life cycle,

DOI: 10.1201/9781003196785-6

from cradle to grave. In the LCA study of photovoltaic technology, the following stages are typically included: raw material and energy supply, production of materials, manufacturing of modules and other parts of the system (inverters, cabling and mounting parts), transport and installation, maintenance (cleaning and repairs), dismantling and transportation, end-of-life scenarios (recycling, incarnation and landfilling) (IEA, 2020). The scheme of the subsequent stages is depicted in Figure 6.1. The balance of system indicated in Figure 6.1 includes inverters, cables, transformators and mounting structure. According to the recommendations of International Energy Agency, the LCA of a given photovoltaic technology should address various life cycle impact indicators such as climate change, ozone depletion, human toxicity (carcinogenic and non-carcinogenic), particulate matter, ionizing radiation, photochemical ozone formation, acidification, eutrophication (terrestrial, freshwater and marine), ecotoxicity, land use, water use and resources use (minerals, metals and fossils). The fundamental input parameters that should be provided to perform LCA of a particular PV technology are the type of the installation (e.g., roof-top and BIPV), efficiency, lifetime and degradation rate of the module, performance ratio of the system, irradiation and expected amount of electric energy production in the chosen location (IEA, 2020).

For assessing the environmental aspects quantitatively, different indicators have to be calculated. One of the important environmental quality indicators usually determined when LCA is applied to PV system is energy payback time (EPBT) – the ratio of the cumulative energy demand (CED), which is the total primary energy input during the entire lifecycle, to the primary energy saving due to annual energy production by the PV module – E_{GEN} (Mustafa et al., 2019):

$$\text{EPBT} = \frac{\text{CED}}{E_{GEN}}. \tag{6.1}$$

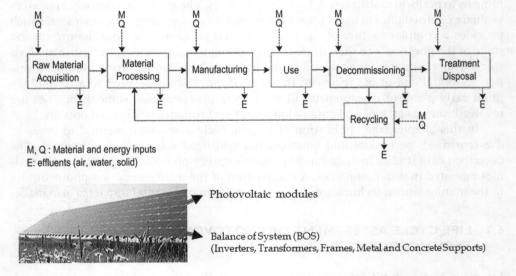

Figure 6.1 Steps of life cycle assessment of photovoltaic technology. (Copied under Creative Commons Attribution 3.0 License from Anctil and Fthenakis (2012).)

The CED expressed in MJ includes the primary energy demand for the production and transport of materials used during the entire lifecycle, fabrication of PV modules, installation of the PV system and end-of-life management. The E_{GEN} can be calculated according to the following equation (IEA, 2020):

$$E_{GEN} = \frac{E_{agen}}{\eta_g} - E_{om} \qquad (6.2)$$

where E_{agen} – the annual electricity production (kWh), η_g – the grid efficiency, which means the primary energy-to-electricity conversion efficiency (kWh/MJ oil-eq), and E_{om} – the annual energy for operation and maintenance (MJ oil-eq).

The next parameter, energy return factor (ERF), which is a ratio of lifetime (L) to EPBT, expresses savings of energy per unit of invested energy:

$$ERF = \frac{L}{EPBT}. \qquad (6.3)$$

Another important environmental quality factor is defined as all greenhouse gas (GHG) emissions expressed in kilograms of CO_2 equivalent per 1 kWh of electric energy generated by the PV modules (g eq-CO_2/kWh). The GHG emission rate can be calculated according to the following formula:

$$GHG = \frac{\text{Total GHG emission during lifecycle}\left(g\,eq\,CO_2\right)}{\text{Annual electricity production}\left(\frac{kWh}{year}\right) \times L(year)}, \qquad (6.4)$$

in which L – the lifetime and GWP – the global warming potential, which assesses the impact of the entire lifecycle on global warming, expressed in terms of equivalent of carbon dioxide (Celik et al., 2016).

The next indicator is the net energy ratio (NER), which can be calculated as the quotient of life cycle energy output and input. When NER indicator is <1, the technology cannot be classified as a sustainable one (Parisi & Basosi, 2015).

6.3 THE END-OF-LIFE TREATMENT OF PHOTOVOLTAIC MODULES

The LCA study should cover the whole life of the given technology, from cradle to grave; thus, the end-of-life treatment of photovoltaic modules is an important issue.

In general, there are a few possible ways of end-of-life treatment of photovoltaic modules: incineration of the components and storage as a hazardous waste, landfilling or recycling based on the procedures adapted from silicon modules, which usually includes 68.8% of glass and 90% of metals (Latunussa et al., 2016). In general, the use of recovered materials is advantageous, as it ensures the supply irrespective of the current market or political problems and can be offered at a lower price. The LCA results show that the utilization of the materials retrieved from old disassembled photovoltaic panels can also lead to savings of energy in the whole loop of the manufacturing process. In case of crystalline silicon modules, the recycling of aluminum and copper brings high benefits.

Recycling is the solution that can reduce the environmental impact of end-of-life dismantled modules; however, in practice, it usually takes place only in the countries where it is profitable and where appropriate legal regulations are implemented. Globally, it is estimated that nowadays only 10% of photovoltaic modules are recycled (Lunardi et al., 2018); thus, the introduction of specific local legislation is crucial. The EU Directive 2012/19/EU of the European Parliament and of the Council on Waste Electrical and Electronic Equipment (WEEE) was extended to photovoltaic products, and the requirement to recycle 80% of materials used in PV modules was introduced (Directive 2012/19/EU).

The possible and expected commercial spread of the third generation photovoltaics will without a doubt involve the necessity of waste management. However, in the LCA studies of emerging PV technologies, the end-of-life phase is most often omitted, mainly due to the lack of real data on dismantled modules, since the new technologies are not adopted in the market. Therefore, the question of recyclability in the case of new technologies still remains open and it is not clear how different options of disposal will affect the life cycle impact on the environment; however, the research suggests that recycling or recovery will be recommended. There are a few prospective options of recycling third generation photocells listed in the order of increasing energy required: reviving/restoring the device (e.g., replenishing the amount of electrolyte in the photoelectrochemical device), reusing a component (e.g., conductive glass substrate), as well as recycling a compound and recovering elements (Miettunen & Santasalo-Aarnio, 2021).

Taking into account that the particular types of third generation solar cells have a similar structure, it is possible to approximately analyze the ways of end-of-life treatment of the individual parts of the photocells.

Conductive glass, i.e., glass substrate coated with a transparent conductive film, is a component used for electrodes in all types of the photocells belonging to the class of third generation. As a relatively expensive material, it has high potential to be reused in photocells with a financial advantage. For example, glass covered with fluorine tin oxide (FTO) could be reused in new dye-sensitized solar cells; however, it requires removing the TiO_2 layer as well as residues of dye and electrolyte. Moreover, such FTO glass can be reused only in the photocells of the same size, since the FTO surface is geometrically limited and the viability of reuse depends on the amount of work and energy involved. Plastic, which is cheap in general and whose recycling in small amounts is not profitable, is employed as a substitution of glass substrates for electrodes in flexible photocells. However, the usage of plastic can be beneficial, since the recovery of Ag which forms a grid for electric contact on the electrodes makes sense only if light substrates, e.g., plastic, are used instead of typical glass, since glass increases the weight ratio to over 100,000 g/ton, which is recognized as the limit of economic profitability (Miettunen & Santasalo-Aarnio, 2021).

Regarding the counter electrodes, the retrieval of their widely used coverage, which is platinum, is viable when it is possible to recover more than the limit of 10–50 g/ton. This can be achieved when a low-weight substrate, such as plastic or extra thin glass, is used. Carbon is a more sustainable material than the metallic coverage of counter electrodes, for it can be obtained from a variety of bio-wastes. The retrieval of other materials comprising the rest of the photocell is complicated, since photoactive layers and charge transport layers are multicomponent and they appear in small amounts, in nanostructural form.

6.4 APPLICATION OF LIFE CYCLE ASSESSMENT TO EMERGING PHOTOVOLTAIC TECHNOLOGIES

The development of a complete assessment of emerging photovoltaic technologies is limited by the uncertainties that result from the lack of production on a large scale, exploitation research or end-of-life pathways in industrial scale. Therefore, the application of the LCA method to the technologies which are under development and have not been tested in real operation conditions remains a challenge. Usually, the LCA studies performed for the third generation photovoltaics are based on laboratory data for single cells or the data from experimental installations of prototype modules. The less popular prospective studies overcome difficulties by considering different scenarios and probability distributions of the analyzed parameters. The overview of different approaches in this respect is presented below, separately for each type of emerging technology.

6.4.1 Life cycle assessment of organic photovoltaic cells

Organic photovoltaic technology is characterized by lower efficiencies and shorter lifetime than mature PV technologies based on monocrystalline and polycrystalline silicon. However, the continuous increase in the OPV efficiency and wide application potential motivated numerous LCA studies, which cover not only laboratory cells, but also roof-top installation, building-integrated modules and cells supplying portable devices.

The early LCA study was performed for the laboratory organic photovoltaic cell exhibiting a PCE of 5%. The detailed analysis of every single step and the material used included the preparation of glass with ITO coverage, PEDOT:PSS hole transport layer and photoactive layer deposition by spin-casting, as well as thermal evaporation of Al and Ca thin-film back electrode. The photoactive layer was a blend of poly-3-hexylthiophene (P3HT) polymer and PCBM fullerene derivative based on C_{60} or C_{70} fullerene. The final energy necessary for the production of OPV cell was estimated as 2800.79 MJ/m^2, which was a similar value to thin-film PV technologies and 50% lower than for Si PV modules in the presented research (Garcia-Valverde et al., 2010). The most burdensome factor was the fabrication of electrodes, especially glass coated with indium tin oxide (ITO), the additional drawback of which is the content of scarce and expensive indium. The estimated energy payback time (EPBT) was 4 years for an assumed efficiency of 5% and 2 years for predicted 10% in the future.

Another approach was presented in the study devoted to the analysis of full roll-to-roll fabrication procedure of an OPV module, which consists of the following steps: sputtering of ITO on the PET substrate, slot-die deposition of ZnO, P3HT:PCBM and PEDOT:PSS as well as screen printing of silver back electrode. The calculations were based on the detailed material inventory, which included the amount of energy necessary for the production of raw materials, consumption of energy in every fabrication process and emission of pollutants (e.g., during transport and recycling process) in the entire lifetime. The LCA study demonstrated that the OPV technology has a low environmental impact with an energy payback time of 1.35 years and ERF 11.12

for 1 m² of the module with 3% efficiency and 67% of active surface. Both EPBT and ERF were dependent on the assumed efficiency and active area of the produced OPV module. The energy for direct processes of module fabrication requires 76.99 MJ/m², and the embodied energy in materials is 302.27 MJ/m² (Espinosa et al., 2011). It is worth noticing that the manufacturing of ITO and silver electrodes accounts for 87% of the embodied energy in all input materials (Figure 6.2) and over 50% energy consumed in the production phases (Figure 6.3).

Since ITO was demonstrated as a burdensome factor, the substitution of ITO with sputtered Al/Cr electrode in a flexible OPV module was investigated and evaluated by the LCA method. The active layer in the studied module consisting of the blend of P3HT:PCBM and PEDOT:PSS was used as a hole transport material, the Ag screen-printed film served as a front electrode, and finally, the polyester foil was employed to laminate the module. The LCA of this promising ITO-free approach was based on the data of energy and material input and output and emission output of the whole roll-to-roll manufacturing process. Energy requirements included the energy needed to produce the raw chemicals and other materials (e.g., sputtering targets) and also the energy utilized in the processes of materials deposition. The consumption of materials included not only constituents of the photocell structure, but also working gas necessary in the sputtering process and adhesive resin used for sealing the foil. The LCA analysis was performed for 1 m² of the processed OPV module on the assumption of realistic performance ratio of 80% for the module working under external conditions. The estimated EPBT decreased from 9.45 years for the lowest active area of module

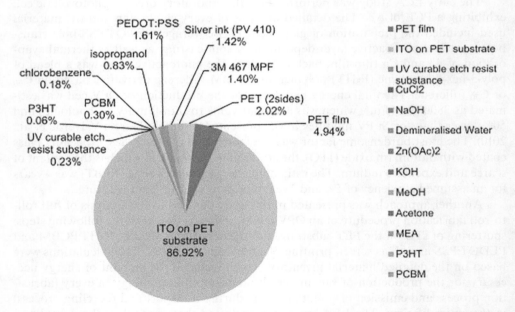

Figure 6.2 Share of energy consumed in the processing of each material during the production of OPV module. (Reprinted from Espinosa et al. (2011). Copyright with permission from Elsevier.)

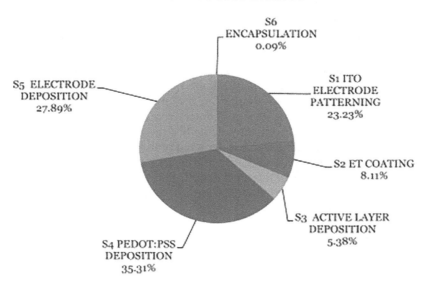

Figure 6.3 Energy directly consumed in the steps of the manufacturing process of OPV module. (Reprinted from Espinosa et al. (2011). Copyright with permission from Elsevier.)

of 36.78% and efficiency of 1% to 0.41 years for the highest active area of 85% and efficiency of 10%. The ERF increased from 1.59 to 36.7 at analogical changes of active area and efficiency. The CO_2 emission factor, which is also the result of the LCA analysis, exhibited the value of 137.68 g eq-CO_2/kWh at PCE of 1% and 55.07 g eq-CO_2/kWh at PCE of 5% (Espinosa et al., 2012).

The interesting LCA study of OPV devoted to single-junction cells based on different photoactive materials and multi-junction devices demonstrated the comparison of environmental quality indicators for different materials and processes used in organic photovoltaics. The results show that the energy embodied during the evaporation of small molecules under vacuum conditions is lower than for solution-processed polymers. However, the highest values of embodied energy are characteristic of fullerenes, since their production is an energy-consuming process. The multi-junction approach is not beneficial for greater complexity of the fabrication and higher consumption of materials; thus, it can be attractive only if the efficiency of the multi-junction device significantly outperforms polymer-based single-junction devices (Anctil & Fthenakis, 2012).

The OPV cell consisting of two cells making tandem cell was the subject of the LCA study in which detailed consumption of materials and energy was taken into account. Both cells consisted of silver electrode, PEDOT:PSS, ZnO, P$_3$HT:PCBM covered with encapsulation material and PET lamination. The energy required to manufacture 1 m^2 of the proposed tandem cell (without the energy embedded in the materials) is 35% higher than for a single-junction device. Assuming a PCE of 3% for tandem OPV cell and 3.5% for single-junction cell, the EPBT was calculated as 0.24 years and 0.17 years, respectively (Espinosa & Krebs, 2014). The life cycle impact assessment indicates that the usage of silver, PET and electricity were the most burdensome factors, accounting for 60%–80% of the most evaluated impacts.

The subject of the LCA research was also another tandem OPV device consisting of a top cell with a high-bandgap photoactive material and a bottom cell with a low-bandgap photoactive material deposited on a flexible PET substrate and encapsulated. The electrode materials that substituted ITO were nano-silver, nano-zinc oxide and PEDOT. According to the simulation, the environmental impacts such as global warming potential, cumulative energy demand, ecotoxicity and metal depletion linked to the manufacturing of 1 m^2 of the tandem OPV module account for 3%–10% of those for multi-Si or CdTe modules, which is depicted in Figure 6.4. Considering the integration of PV modules with a facade of a building, the calculations show that the EPBT for OPV is 18%–55% of the EPBT for the multi-Si and CdTe technologies, whereas the global warming potential is around 50% lower than for the considered reference modules (Hengevoss et al., 2016).

The valuable research work concerned also juxtaposition of OPV with the mature silicon PV technology in terms of life cycle impact. Interesting results were brought by the comparison of organic photocells based on various materials (all-polymer technology and bulk heterojunction of polymer and fullerene derivative) with monocrystalline Si and amorphous Si cells. Two different systems were examined: roof-top array with 25 years lifetime and portable charger with a short lifetime of 5 years. Additionally, incineration and landfilling were considered as two possible options for the end-of-life scenario, as well as balance of system (mounting structure, inverter and cables) was included in all analyzed variants (Tsang et al., 2016). Taking into account all studied options, the results of LCA exhibited the environmental impact of OPV was 55% lower for roof-top installation and 70% lower for the charger, compared to the Si technology. Assuming a PCE of 5% for each OPV variant, 13% for mono-Si and 6.5% for a-Si, the

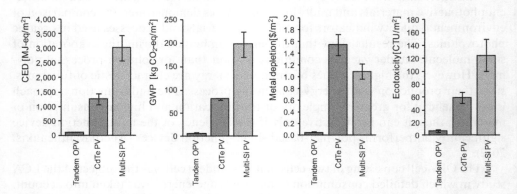

Figure 6.4 Environmental impact of the production of 1 m^2 of flexible tandem OPV module, compared with CdTe and multi-Si technologies, expressed in the following categories: CED, GWP, metal depletion and ecotoxicity. (Reprinted from Hengevoss et al. (2016). Copyright with permission from Elsevier.)

estimated energy payback time changes from 42 to 449 days for OPV and from 492 to 919 days for the Si PV technologies (Tsang et al., 2016).

In the context of life cycle assessment, the advantage of OPV over typical silicon cells was also proven in another broad analysis, which besides the device type and architecture included lifetime, transportation, efficiency degradation in time, disposal, cost and the projected technological progress. The following assumptions were made for the OPV technology: 8.7% as the current value of efficiency and 15% as a projected value, lifetime 5 years, degradation 1.38% per year and the price 0.28 $/Wp. For the silicon module to which the OPV technology was compared, the specifications used for the calculations were current efficiency of 19.9% and 23% in the future, 25 years lifetime, degradation of 0.91% per year and the price 0.37$/Wp (Krebs-Moberg et al., 2021). Regarding the recycling issue, in the case of OPV, only PET foil recovery of 100% was assumed; however, for the Si module, the recovery of over 90% glass and aluminum was taken in the calculations. The research showed lower environmental impacts of OPV compared to the Si modules in all studied categories, independently of the assumption of recycling. Even when landfilling is considered, the impact of organic modules remains the lowest in comparison with the Si technology; however, the research on the landfilling scenario indicates that it is associated with the risk of soil and ground water contamination (Espinosa et al., 2016). The experimental study of the Ag and Zn emissions from the OPV modules operating under real outdoor conditions showed that properly encapsulated and sealed modules are not a threat for the environment; however, significant emissions of both metals were observed from the damaged modules.

6.4.2 Life cycle assessment of dye-sensitized photovoltaic cells

Another representative of third generation PV, i.e., dye-sensitized solar cells, was also studied in the frames of the LCA methodology. The subject of investigation was the dye cell consisting of glass with transparent conductive oxide, titanium dioxide, organic dye trimethylamine, iodine-based electrolyte platinum, silver paste, as well as polyethylene and polyester resin, made in the glass-glass technology. The manufacturing steps, such as laser scribing, drilling of holes, screen printing and sintering of TiO_2, application of dye and electrolyte, as well as sealing of the cell, were taken into account. The balance of system included an aluminum frame and steel structure for mounting of roof-top-integrated on-grid installation. The results of sustainability evaluation showed that the variation of DSSC efficiency from 6% to 13% implies the change of EPBT in the range of 2.11–0.97 years. If it is assumed that the lifetime is 20 years, the net energy ratio changes from 9.5 to 20.57 and the GHG emission varies from 29.72 to 13.72 g eq-CO_2/kWh (Parisi & Basosi, 2015). When DSSC is compared with other PV technologies, the embedded energy is much lesser than that for c-Si and similar to that for thin films. It was found that in the case of DSSC, the most energy-intense process is the fabrication of glass with conductive coverage.

The broad LCA analysis of DSSC, which started with the production of raw materials (e.g., the synthesis of the sensitizer) and industrial production processes and included the virtual operation of roof-top installation, was also reported. The research performed for three different sensitizers (N719 ruthenium dye, D5 dye and YD2-o-C8 dye) proved that solar FTO glass exhibits the major impact in all environmental categories (Parisi et al., 2014). The comparison of DSSC in a few variants with other PV technologies (based on organic polymer, MCPH – micromorph tandem of a-Si cell on microcrystalline Si cell, CdTe, CIS and a-Si) in terms of cumulative energy demand (Figure 6.5) indicates that flexible DSSC on PET substrate can be competitive to more mature technologies, especially if integration with architecture without steel frame is considered. Among the studied PV technologies, the DSSC module on PET exhibits the lowest EPBT of 0.73 years and below 1.5 years when BOS is included. The global warming potential indicator achieves the best values for a micromorph Si module, which is followed by DSSC on PET; again, this indicates the advantage of flexible DSSC over other variants (Parisi et al., 2014).

The described roof-top applications of DSSC are not anticipated to be popular in the future, whereas different niches of the photovoltaic market, such as indoor applications and building-integrated photovoltaics (BIPV), will be available for this very technology, owing to easy work in low-intensity and diffuse light, as well as semi-transparency and the possibility to control the color of the photocells. Therefore, it is important to take into account such solutions in the assessment of the environmental impact performed using the LCA methodology.

A detailed LCA research on window-integrated PV system placed on the building facade consisting of DSSC modules was carried out under Malaysian climate. The assumed energy consumption in the complete process of fabrication and assembly of a DSSC module included drying of subsequent layers of deposited materials, cleaning excess of dye and ultrasonic soldering, in addition to main steps such as TiO_2 printing or electrolyte filling. It is worth noticing that the comparison of the energy consumed

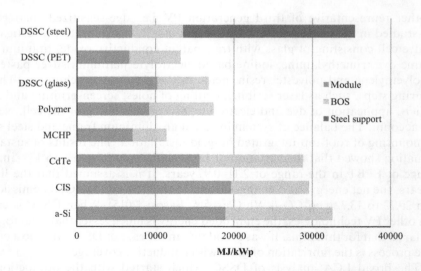

Figure 6.5 The CED indicator for different considered photovoltaic technologies including modules, BOS and, additionally, steel support in case of dye cells on steel substrate – DSSC (steel). (Reprinted from Parisi et al. (2014). Copyright with permission from Elsevier.)

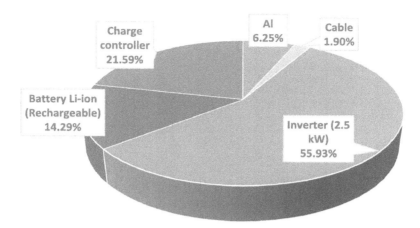

Figure 6.6 Detailed energy consumption for the BOS of DSSC. (Reprinted from Mustafa et al. (2019). Copyright with permission from Elsevier.)

in different production processes indicates drying of printed TiO_2 film (345.6 MJ/m^2) and staining in dye (103.68 MJ/m^2) as the most energy-intense stages. In energy consumption for BOS, the inverter is responsible for over 50% (Figure 6.6). In terms of embodied energy of input materials, the largest contribution of FTO reported earlier has been confirmed. In the total energy embodied in input materials for the considered 18.52 m^2 of DSSC modules, the FTO glass accounts for 73.09% due to the large amount of energy consumed in glass production and large mass that impacts the transportation (Mustafa et al., 2019). Platinum with 18.93% share and EVA foil with 3.97% share occupy the next two places in percentage distribution on embodied energy. The CED expressed in MJ/kWh for different stages of production shows that the manufacturing of module and panel is the main contributor and exceeds the sum of other three factors including BOS, DSSC operation, and maintenance and transportation. The GHG emissions represent an analogous trend, which is depicted in Figure 6.7.

On the basis of the assumption of DSSC module with a PCE of 5% and 20 years lifetime, the estimated EPBT was 2.42 years and the total GHG emissions were estimated as 70.52 g eq-CO_2/kWh. The relatively high values of these indicators are due to the lower PCE of the analyzed module than in other studies.

As a technology attractive for future applications in BIPV, dye-sensitized solar modules were also studied from the point view of environmental impact in a long-term LCA approach. Different scenarios based on various stages of improvement of DSSC technology (e.g., efficiency of module changing in the range of 3.91%–6.25% and life time 5–15 years) were considered, and the environmental impact was projected to the years 2021 and 2050 (Parisi et al., 2020). The outcome of such an analysis provided CED decreasing from 6.0 to even 0.9 MJ/kWh for the assumed drop of embodied energy for raw materials and energy consumed during the fabrication process. The EPBT was estimated as 2.3–1.1 years for the same scenarios and GHG emissions as 350 to 50 g eq-CO_2/kWh. Silver and ruthenium indicated in this work as the materials

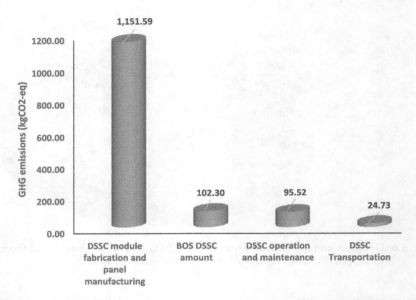

Figure 6.7 GHG emissions for fabrication, operation, maintenance and transportation of DSSC and BOS. (Reprinted from Mustafa et al. (2019). Copyright with permission from Elsevier.)

responsible for the highest environmental footprint should be substituted with more eco-friendly replacements. Therefore, the search for alternative materials, especially the development of new metal-free sensitizers, is a pivotal issue from the LCA perspective. In order to reduce the energy consumed in glass production and partially solve the problem of large contribution of the FTO glass in the indicators, which was also addressed in earlier reports, the usage of much thinner glass was recommended. Thin flexible glass is more expensive; however, it is suitable for low-cost roll-to-roll manufacturing of the cells. The juxtaposition of DSSC with traditional silicon and thin-film technologies in terms of EPBT revealed the comparable values for these technologies, which proves an excellent performance of the DSSC used in BIPV.

6.4.3 Life cycle assessment of perovskite photovoltaic cells

Perovskite solar cells demonstrate a rapid improvement in the photoconversion efficiency and great potential for commercialization; thus, despite being a relatively new technology, they are also a subject of LCA analysis.

The results of the LCA study on perovskite photocells depend mainly on the differences in the structure of the cells and deposition methods of perovskites. That was demonstrated in the multi-variant investigation, in which the HTL-free cell and the structure with CuSCN as HTL, as well as the vapor-based deposition and solution-based deposition of perovskite were considered. The summary of the results in nine impact categories determined for the manufacturing of 1 m^2 of perovskite module placed the HTL-free perovskite module between first generation PV (mono-Si and poly-Si) and second generation PV (a-Si, CdTe and CIS) in terms of environmental impact. Taking into account impacts per 1 kWh of generated electricity, higher values were obtained for perovskites than for the reference commercial Si technologies. The assumption of 15% PCE and relatively short lifetime of 5 years led to EPBT in the range of 1.05–1.54 years and GWP 100–150 g eq-CO_2/kWh (Celik et al., 2016). The major contribution (50%–90%) in most categories was from electricity consumption during the production of the photocell. Two deposition methods applied to obtain the perovskite layer were considered. The deposition of lead perovskite absorber by co-evaporation, which requires vacuum conditions, was recognized as the most energy-intense process. The co-evaporation method had twice as much share as the spray deposition from solution. The structure of the perovskite cell in which the HTL growth was eliminated proved the potential to decrease material cost and energy requirement by the introduction of carbon back contact. In the presented cradle-to-gate LCA study, the impacts linked to the toxicity of lead occurring in perovskite material were negligible; however, a full analysis including the end-of-life stage is necessary in this respect.

From the point of view of future commercialization development, the usage of lead in perovskite cells raises a serious environmental concern. Since lead is harmful and stable in the ecosystem, both manufacturing and end-of-life stages potentially can pose a risk to humans and cause long-term eco-system damage (Bae et al., 2019). However, the LCA analysis devoted to the comparison of the photocells based on lead perovskites and lead-free perovskites, as well as all-inorganic lead perovskite showed that, among the nine studied environmental categories, the share of the perovskite layer fabrication in human toxicity potential (HTP) is from 0.29% for $MAPbI_2Cl$ to 9.74% for $FAPbI_3$ perovskite material and the main reason is the usage of isopropanol to clean the perovskite film. The considered perovskite cells, with the determined manufacturing energy 202–254 kWh/m^2 and EPBT 0.9–3.5 years, provided better results than Si cells; however, the GHG emissions (2553–6466 g eq-CO_2/kWh) are higher than for other PV technologies (Zhang et al., 2017; Ludin et al., 2018). The assumed lifetime of the photocells was 1 year and conversion efficiency 4.88%–20%, depending on the type of perovskite material. The role of lead salts used in the manufacturing process of perovskite reflected by 0.13%–0.22% share of HTP is negligible due to the small amount of the salts used in the synthesis of perovskite. Thus, the environmental impact of a heavy metal depends mainly on the end-of-life scenario. In this respect, incineration of dismantled modules with energy and steam recovery was indicated as beneficial comparing to landfilling. The minor role of Pb in the context of life cycle assessment of perovskite photocells was also proven in another broad study which included two types of cell architecture, three different perovskite deposition methods and three end-of-life scenarios (Alberola-Borràs et al., 2018). The investigations showed that recycling and reuse of lead from perovskites does not significantly improve environmental indicators.

An important factor considered in the LCA research on PSC was also the origin of lead used in the perovskite layer. Taking into account five different toxicity indicators, it was demonstrated that the Pb derived from recycling is more beneficial than the Pb from concentrate or recycled Pb mixed with the one from concentrate (Ibn-Mohammed et al., 2017).

Among the raw materials used in PSC manufacturing, gold was found to have the largest environmental impact, since both gold extraction and the deposition of gold layer by thermal evaporation are energy-intense processes. Additionally, the extraction of gold involves the release of toxic chemicals, which affects the eco-indicators. The usage of Au substitutions, such as aluminum or silver, can significantly improve the environmental profile of back electrode material in PSC (Zhang et al., 2017; Maranghi et al., 2019). Other materials that burden the environmental footprint of PSC are conductive solar glass and TiO_2. In the study performed according to the scheme depicted in Figure 6.8, conductor deposition, which represents the Au cathode, accounts for 52% of the energy embodied in materials and 19% of electrical energy usage for deposition. Substrate patterning, which corresponds to the FTO glass, accounts for 44% of the energy embodied in materials and for 28% of energy for cleaning, printing and sintering (Ibn-Mohammed et al., 2017). Deposition of perovskite layer is responsible for 25% of energy used for processes; however, the presented data correspond to energy-intense vapor deposition, instead of the less demanding spin coating, for example.

The LCA analysis carried out separately for the typical part of PSC device (ETM, HTM and contacts) and perovskite layer, which was deposited by different methods, showed that common parts have a predominant impact on the obtained results. The study based on laboratory data of the materials and energy used in the preparation

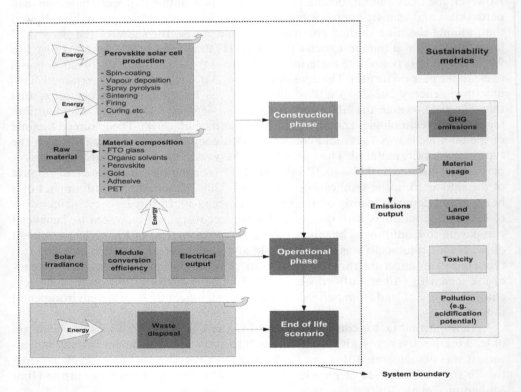

Figure 6.8 System boundary considered in the LCA study on perovskite solar cells. (Copied under Creative Commons CC-BY-NC-ND license from Ibn-Mohammed et al. (2017).)

of perovskite solar device indicates great importance of energy consumption in the results of LCA. Attention is drawn to the fact that the synthesis of the perovskite layer, which is the essential part of the cell, is much less intense than the deposition of all other typical layers. The share of energy consumption in cumulative energy demand is a few orders of magnitude higher in all of the categories, for all analyzed variants of PSC (Alberola-Borràs et al., 2018). The same investigation included different end-of-life scenarios for PSC: residual landfill, reuse and residual landfill and reuse together with recycling. The results showed that reuse and recycling has a positive impact on toxicity indicators. However, in general, disposal has low contribution in total cumulative energy demand estimated for different configurations of PSC.

A recent broad research on the environmental impact of perovskite cell was based on the detailed inventory of materials and energy consumption, as well as assumptions of 11.6% PCE (an increase of up to 23% was projected due to the development of technology), lifetime of 5 years and degradation of 1.38%/year. The following parts constituted the PSC structure: ITO-coated glass substrate, ZnO ETL, lead perovskite, spiro-OMeTAD HTL, Ag contact and PET lamination. In this architecture, ITO and Ag were employed as the front and back electrodes, instead of FTO and Au; nevertheless, these two components still burden the environment. The evaporation of Ag cathode takes over twice as much energy as all other processes, silver accounts for 47.3% share in mineral resource scarcity category, and the contribution of ITO glass is over 70% in 10 impact categories among all 14 evaluated impact categories (Krebs-Moberg et al., 2021). Regarding the end-of-life scenarios, recycling is more beneficial in comparison with landfilling; particularly, the reuse of ITO-coated glass substrates is recommended due to the expected decrease in the ecotoxicity indicator. The comparison of LCA results for perovskite and silicon PV technology showed that PSC is definitely less environmentally burdensome. It was demonstrated that in 13 out of 14 impact categories, PSC is responsible for 25%–45% of silicon technology contribution.

6.4.4 Life cycle assessment of quantum dot-sensitized photovoltaic cells

The number of LCA studies devoted to quantum dot-sensitized solar cells is relatively low, taking into account that the efficiency of this technology has significantly improved in recent years, making it a promising solution. The latest achievement of 15.31% efficiency was accomplished due to the modification of the QDs deposition method, which enables us to increase the load of the dots on the TiO_2 electrode (Song et al., 2021). Besides the improvement in the photovoltaic parameters of QDSSC, it is worth emphasizing that non-toxic QDs containing Zn, Cu, In, S and Se were employed; however, in most of the early reports, the quantum dots based on Pb and Cd were used, which arouse the concerns due to toxicity of these elements.

The LCA method was also applied to hybrid QD-based photocells prepared in laboratory scale, with the assumption of 10% efficiency, 75% performance ratio and 25 years of lifetime without consideration of end-of-life scenarios. On the basis of these values and the expected energy output of 31.88 kWh/Wp, very low values of EPBT of 1.51 years, GHG of 2.89 g eq-CO_2/kWh and CED of 370 MJ/m^2 were obtained (Azzopardi & Mutale, 2010). The simulation of efficiency and lifetime influence on the net energy ratio (NER) showed that the minimum 2 years of lifetime and 8% efficiency should enable us to reach sustainability of QDSSC.

More optimistic assumptions (14% efficiency, 80% performance ratio and 25–30 years of lifetime) made in the analysis which included the life phases from raw materials acquisition to exploitation of the QDSSC module provided EPBT 0.9 years, GHG 5 g eq-CO_2/kWh and CED 1029.6 MJ/m^2 (Şengül & Theis, 2011). The comparison with other renewable and non-renewable energy sources provided in that report led to the conclusion that the environmental impact of the QDSSC modules is lower in all categories than that of carbon-based energy sources and the EPBT of QDSSC is longer than for wind technology.

6.5 SUMMARY

The comparison of the LCA results obtained for particular emerging photovoltaic technologies is hampered mainly due to the discrepancies in the assumptions of efficiency and lifetime of the photocells made at the early stage of the analysis. The values of the determined environmental indicators are also influenced by the type of the materials that constitute the structure of the given photocell and the type of the considered installation (e.g., roof-top, portable device, BIPV or single laboratory cells).

Nevertheless, the overview of most LCA reports indicates electrode materials as the important burdensome factor. The production and deposition of ITO, FTO, Au, Ag on glass or foil materials are energy intense and greatly affect the environment. Therefore, the substitution of these electrode materials with less expensive, abundant alternatives and recycling of the used electrodes is recommended; however, recycling involves a few stages that also need energy and chemicals. The remaining parts of the cells, especially the photoactive layer and electron transport material, are composed of nanomaterials that are used in small amounts and thus are difficult to recover and reuse.

The consideration of the main LCA indicators, such as CED, EPBT and GHG emissions, leads to the important conclusion that the emerging technologies exhibit a lower environmental impact than the most widely used silicon solar cells. However, the attention is drawn by the relatively high GHG of the perovskite technology, which is the most promising one compared to OPV, DSSC and QDSSC, for the highest efficiency achieved. Another environmental issue that has to be addressed in the case of perovskites is also the necessity of lead substitution with non-toxic metals (e.g., Sn, Cu and Bi), especially if landfilling of modules is an end-of-life option.

REFERENCES

Alberola-Borràs, J.A., Vidal, R., Juárez-Pérez, E.J., Elena Mas-Marzá, E., Guerrero, A. & Mora-Seró, I. (2018) Relative impacts of methylammonium lead triiodide perovskite solar cells based on life cycle assessment. *Solar Energy Materials and Solar Cells* 179, 169–177. doi: 10.1016/j.solmat.2017.11.008.

Anctil, A. & Fthenakis, V. (2012) Life cycle assessment of organic photovoltaics. In: Fthenakis, V. (ed.), *Third Generation Photovoltaics*. IntechOpen: London. doi: 10.5772/38977; https://www.intechopen.com/chapters/32591.

Azzopardi, B. & Mutale, J. (2010) Life cycle analysis for future photovoltaic systems using hybrid solar cells. *Renewable and Sustainable Energy Reviews* 14(3), 1130–1134. doi: 10.1016/j.rser.2009.10.016.

Bae, S.Y., Lee, S.Y., Kim, J.W., Umh, H.N., Jeong, J., Bae, S., Yi, J., Kim, Y. & Choi, J. (2019). Hazard potential of perovskite solar cell technology for potential implementation of "safe-by-design" approach. *Scientific Reports* 9(1), 1–9. https://www.nature.com/articles/s41598-018-37229-8.

Celik, I., Song, Z., Cimaroli, A.J., Yan, Y., Heben, M.J. & Apul, A. (2016) Life Cycle Assessment (LCA) of perovskite PV cells projected from lab to fab. *Solar Energy Materials and Solar Cells* 156, 157–169. doi: 10.1016/j.solmat.2016.04.037.

Directive 2012/19/EU of the European Parliament and of the Council of 4 July 2012 on waste electrical and electronic equipment (WEEE). Available from: https://eur-lex.europa.eu/legal-content/EN/TXT/?uri=celex%3A32012L0019 [Accessed December 15, 2021].

Espinosa, N. & Krebs, F.C. (2014) Life cycle analysis of organic tandem solar cells: when are they warranted? *Solar Energy Materials and Solar Cells* 120, 692–700. doi: 10.1016/j.solmat.2013.09.013.

Espinosa, N., García-Valverde, R., Urbina, A. & Krebs, F.C. (2011) A life cycle analysis of polymer solar cell modules prepared using roll-to-roll methods under ambient conditions. *Solar Energy Materials and Solar Cells* 95(5), 1293–1302. doi: 10.1016/j.solmat.2010.08.020.

Espinosa, N., García-Valverde, R., Urbina, A., Lenzmann, F., Manceau, M., Angmo, D. & Krebs, F.C. (2012) Life cycle assessment of ITO-free flexible polymer solar cells prepared by roll-to-roll coating and printing. *Solar Energy Materials and Solar Cells* 97, 3–13. doi: 10.1016/j.solmat.2011.09.048.

Espinosa, N., Zimmermann, Y.-S., dos Reis Benatto, G.A., Lenz, M. & Krebs, F.C. (2016) Outdoor fate and environmental impact of polymer solar cells through leaching and emission to rainwater and soil. *Energy and Environmental Science* 9, 1674. doi: 10.1039/c6ee00578k.

Garcia-Valverde, R., Cherni, J.A. & Urbina, A. (2010) Life cycle analysis of organic photovoltaic technologies. *Progress in Photovoltaics* 18, 535–538. doi: 10.1002/pip.967.

Hengevoss, D., Baumgartner, C., Nisato, G. & Hugi, C. (2016) Life cycle assessment and eco-efficiency of prospective, flexible, tandem organic photovoltaic module. *Solar Energy* 137, 317–327. doi: 10.1016/j.solener.2016.08.025.

Ibn-Mohammed, T., Koh, S.C.L., Reaney, I.M., Acquaye, A., Schileo, G., Mustapha, K.B. & Greenough, R. (2017) Perovskite solar cells: an integrated hybrid lifecycle assessment and review in comparison with other photovoltaic technologies. *Renewable and Sustainable Energy Reviews* 80, 1321–1344. doi: 10.1016/j.rser.2017.05.095.

IEA (2020) Methodology Guidelines on Life Cycle Assessment of Photovoltaics. Available from: https://iea-pvps.org/key-topics/methodology-guidelines-on-life-cycle-assessment-of-photovoltaic-2020/ [Accessed September 10, 2021].

Krebs-Moberg, M., Pitz, M., Dorsette, T.L. & Gheewala, S.H. (2021) Third generation of photovoltaic panels: a life cycle assessment. *Renewable Energy* 164, 556–565. doi: 10.1016/j.renene.2020.09.054.

Latunussa, C.E.L., Ardente, F., Blengini, G.A. & Mancini, L. (2016) Life cycle assessment of an innovative recycling process for crystalline silicon photovoltaic panels. *Solar Energy Materials and Solar Cells* 156, 101–111. doi: 10.1016/j.solmat.2016.03.020.

Ludin, N.A., Mustafa, N.I., Hanafiah, M.M. Ibrahim, M.A., Teridi, M.A.M., Sepeai, S., Zaharim, A. & Sopian, K. (2018) Prospects of life cycle assessment of renewable energy from solar photovoltaic technologies: a review. *Renewable and Sustainable Energy Reviews* 96, 11–28. doi: 10.1016/j.rser.2018.07.048.

Lunardi, M.M., Alvarez-Gaitan, J.P. & Corkish, J.I.B.R. (2018) A review of recycling processes for photovoltaic modules. In: Zaidi, B. (ed.), *Solar Panels and Photovoltaic Materials*. IntechOpen: London. doi: 10.5772/intechopen.74390.

Maranghi, S., Parisi, M.L., Basosi, R. & Sinicropi, A. (2019) Environmental profile of the manufacturing process of perovskite photovoltaics: harmonization of life cycle assessment studies. *Energies* 12, 3746. doi: 10.3390/en12193746.

Miettunen, K. & Santasalo-Aarnio, A. (2021) Eco-design for dye solar cells: from hazardous waste to profitable recovery. *Journal of Cleaner Production* 320, 128743. doi: 10.1016/j.jclepro.2021.128743.

Mustafa, N.I., Ludin, N.A., Mohamed, N.M., Ibrahim, M.A., Teridi, M.A.M., Sepeai, S., Zaharim, A. & Sopian, K. (2019) Environmental performance of window-integrated systems using dye-sensitised solar module technology in Malaysia. *Solar Energy* 187, 379–392. doi: 10.1016/j.solener.2019.05.059.

Parisi, M. & Basosi, R. (2015) Environmental life cycle analysis of nonconventional thin-film photovoltaics: the case of dye-sensitized solar devices. In: Reddy, B. & Ulgiati, S. (eds.), *Energy Security and Development*. Springer: Berlin, Germany. doi: 10.1007/978-81-322-2065-7_12.

Parisi, M.L., Maranghi, S. & Basosi, R. (2014) The evolution of the dye sensitized solar cells from Grätzel prototype to up-scaled solar applications: a life cycle assessment approach. *Renewable and Sustainable Energy Reviews* 39, 124–138. doi: 10.1016/j.rser.2014.07.079.

Parisi, M.L., Vesce, S.M.L., Sinicropi, A., Di Carlo, A. & Basosi, R. (2020) Prospective life cycle assessment of third-generation photovoltaics at the pre-industrial scale: a long-term scenario approach. *Renewable and Sustainable Energy Reviews* 121, 109703. doi: 10.1016/j.rser.2020.109703.

Şengül, H. & Theis, T.L. (2011) An environmental impact assessment of quantum dot photovoltaics (QDPV) from raw material acquisition through use. *Journal of Cleaner Production* 19(1), 21–31. doi: 10.1016/j.jclepro.2010.08.010.

Song, H., Lin, Y., Zhang, Z., Rao, H., Wang, W., Fang, Y., Pan, Z. & Zhong, X. (2021) Improving the efficiency of quantum dot sensitized solar cells beyond 15% via secondary deposition. *Journal of American Chemical Society* 143(12), 4790–4800. doi: 10.1021/jacs.1c01214.

Tsang, M.P., Sonnemann, G.W. & Bassani, D.M. (2016) Life-cycle assessment of cradle-to-grave opportunities and environmental impacts of organic photovoltaic solar panels compared to conventional technologies. *Solar Energy Materials and Solar Cells* 156, 37–48. doi: 10.1016/j.solmat.2016.04.024.

Zhang, J., Gao, X., Deng, Y., Zha, Y. & Yuan, C. (2017) Comparison of life cycle environmental impacts of different perovskite solar cell systems. *Solar Energy Materials and Solar Cells* 166, 9–17. doi: 10.1016/j.solmat.2017.03.008.

Chapter 7

Applications of Emerging Photovoltaics – Future Outlook

7.1 INTRODUCTION

Third generation solar cells have drawn the attention of many research groups in recent years. In comparison with traditional silicon photovoltaic cells, the emerging technologies revealed significant advantages, such as low-cost production, low harmfulness to the environment and easy manufacturing procedures. Additionally, third generation photovoltaic modules are constituted of thin layers and are therefore lightweight, as well as compatible with flexible substrates. Semitransparency and variety of colors presented by modern photovoltaics are also of great importance when it comes to applications. Pursuit of commercialization is an obvious direction driven by the promising laboratory performance of the emerging photocells and all desirable attributes they exhibit.

According to the recent data presented in Table 7.1, perovskite-based photovoltaic devices deliver the highest efficiency among the third generation photocells, which is close to the efficiency provided by the Si technology. The perovskite technology has demonstrated the remarkable improvement in recent years and therefore shows great potential for future applications. Table 7.1 presents the photovoltaic parameters for third generation cells and minimodules measured under STC and confirmed by

Table 7.1 Photovoltaic parameters of silicon technology and third generation solar cells

Technology	J_{SC} (mA/cm^2)	V_{OC} (V)	FF (%)	PCE (%)
Si crystalline cell	42.65	0.738	84.9	26.7
Si thin-film minimodule	29.7	0.492	72.1	10.5
Organic cell	24.24	0.8467	74.3	15.2
Organic minimodule	24.48	0.8276	69.6	14.1
Dye cell	22.47	0.744	71.2	11.9
Dye minimodule	20.19	0.754	69.9	10.7
Perovskite cell	22.73	1.178	84.4	22.6
Perovskite minimodule	23.4	1.149	79.6	21.4
Perovskite/Si tandem cell	20.26	1.884	77.3	29.5
Perovskite/CIGS tandem cell	19.24	1.768	72.9	24.2
Perovskite/perovskite tandem cell	16.54	2.048	77.9	26.4
Perovskite/perovskite tandem minimodule	14.22	2.009	75.9	21.7

Source: Data from Green et al. (2021a).

DOI: 10.1201/9781003196785-7

certified laboratories. For comparison, the data for selected multi-junction cells and the first generation Si technology are also shown. Attention is drawn by tandem cells based on perovskite, which stand out due to their great efficiency.

In spite of all the distinctive properties, the third generation solar cells have not reached technological maturity, since they suffer from stability issues, short lifetime and still insufficient efficiency. This chapter summarizes the current state and advancement of the efforts aiming at meeting the requirements of the PV market and search for some niche applications. The subsections are devoted to particular types of third generation photovoltaic technologies, such as organic, dye-sensitized, perovskite and quantum dot-sensitized solar cells, as well as the indoor applications of emerging photovoltaics and tandem structures that bring together different materials.

7.2 ORGANIC PHOTOCELLS

The OPV technology offers flexible, large-area, low-cost modules, which can be semitransparent and visually attractive. Despite these significant advantages, the OPV has not been commercialized so far due to too low efficiency of modules and degradation in external environment; however, some attempts were made in this area.

In general, the performance of photocells under real outdoor conditions is strongly influenced by fluctuations of light intensity, temperature, wind and humidity. Organic materials are intrinsically more vulnerable to degradation from water and oxygen than inorganic compounds. In this context, the proper choice of photoactive organic material, e.g., P3HT and P3CT, which is less prone to light degradation, as well as reliable encapsulation of the cells, is a crucial issue. The research shows that the application of highly resistant encapsulant film and edge-sealing adhesive can allow maintaining the efficiency of OPV modules for at least 3 years (Weerasinghe et al., 2016; Lee et al., 2021). However, despite promising investigations, there is still much room for development of reliable protective coatings that could be applied on industry scale.

The attempts to commercialize the OPV technology include the production of OPV flexible modules by Konarka Technologies®, which started in 2008. The company manufactured light, colorful polymer-fullerene modules called Power Plastic, which provided 3%–5% PCE and found application, for example, as decorative bus stop roofing (Wikipedia, 2022). Unfortunately, in 2012 Konarka ceased the operation due to financial problems.

In 2010, the upscaling of organic cells to an industrial level, including all necessary materials and instrumentation as well as a complete cost analysis, was presented (Krebs et al., 2010). The OPV fabricated in a roll-to-roll process exhibited the following photovoltaic performance parameters: I_{SC} from 60.1 to 185.8 mA, V_{OC} from 7.56 to 7.83 V, FF from 37.9% to 36.7% and PCE from 1.79% to 1.18% for a small-sized module (96 cm^2) and a large-sized module (360 cm^2), respectively. The listed performance parameters were averaged over the results of a few hundreds of the tested modules. The fabricated OPV modules were installed in a few outdoor demonstration projects, but they were not mass-manufactured.

Due to semitransparency, OPV modules are suitable for building-integrated photovoltaics (BIPV) applications. The roll-to-roll manufactured test modules with 10% transparency were integrated into window glass, showing that visually attractive modules can be applied in buildings. These modules consisted of 19 cells connected in series deposited by slot-die coating on PET and exhibited a PCE of 4.3%. Slightly better results were obtained for modules built of 30 cells on glass, which demonstrated a PCE of 4.8% (Lucera et al., 2017).

Regardless of fabrication tests, the laboratory research continuously provides interesting new solutions that can potentially be applied in industrial scale. For example, it is worth paying attention to the novel, promising method of depositing the layer of organic semiconductor for OPV applications. The innovative approach demonstrated that painting by hand with a felt-tip marker containing the solution of semiconductor provides surprisingly good results. The developed deposition method is simple, is of low cost and does not waste much material, as opposed to the usually used spin coating. The layers of organic semiconductor blend P3HT:$PC_{61}BM$ or PTB7:$PC_{71}BM$ covered the glass with ITO/ZnO or glass with an ITO/PEDOT:PSS layer. The efficiency of solar energy conversion reached 3.6% for the pen-coated semiconductor layer of 201 nm thickness and 3.7% for spin-coated 206 nm thickness (Suzuki et al., 2017). The photovoltaic performance of spin-coated and pen-coated OPV cells was very similar, which indicates great potential of the simple deposition method.

The prospective applications of OPV technology include mainly BIPV, powering of low-consuming indoor electronic devices (v.i.) and deposition of light, flexible modules on fabrics (called smart fabrics or wearable technology) (Anctil & Fthenakis, 2012).

7.3 DYE-SENSITIZED PHOTOCELLS

DSSC present many attractive features, such as cost-effectiveness, lightweight of modules and easy fabrication; however, this technology faces several challenges that need to be addressed on the way to wide commercialization. Currently, the improvement of efficiency and long-term stability are the main concerns that hinder the large-scale introduction of DSSC to the market (Mozaffari et al., 2017); however, some companies manufacture DSSC modules for niche applications. The world leader in DSSC production is G24 Power®, which manufactures lightweight, flexible modules for both indoor and outdoor use. As a result of the two-stage process, the 15 cm wide rolls of the length up to 500 m are produced and the size of modules can be suited to the requirements of the customer (G24 Manufacturing Process, 2022).

Experimental installation that enables testing of semitransparent dye-sensitized photovoltaic modules under real weather conditions were mounted, e.g., in Italy and Switzerland. In Italy, semitransparent DSSC modules were prepared by screen printing and connected into panels tested inside a greenhouse. The 35% transparency of the fabricated modules was accompanied by 3.9% PCE with the aperture area of module equal to 312.9 cm^2 (Barichello et al., 2021). The application of DSSC modules in a greenhouse demonstrated the effect of light filtering due to partial absorption of sunlight by the utilized dye, which, however, was selected not to interfere with the photosynthesis. The tests of panels with the active area of 886 cm^2 provided maximum power from 0.22 W for horizontal position of modules under cloudy weather to 2.7 W for 45° tilted modules under sunny conditions.

In Switzerland, the unique large-area DSSC BIPV installation was demonstrated by Solaronix® in the Swiss Tech Convention Center. The multicolor modules of different sizes mounted on southwest glass facade have many functions: they generate electric energy, are decorative, shade and prevent the large hall from overheating (Solaronix, 2022). Other applications of DSSC developed by Solaronix include outdoor or indoor autonomous furniture (e.g., benches) with integrated PV modules which can power portable electronic devices.

Regarding the integration of photovoltaic modules, an interesting prospective approach would also be the usage of colorless transparent modules. Recently, the experimental research has demonstrated near-infrared dye-sensitized solar cells (NIR-DSSC), which are see-through devices based on the cyanine dye absorbing beyond 800 nm. The photocells exhibiting over 75% mean transmittance of visible light work well with both iodide- and cobalt-based electrolytes and provide a PCE of 3.1% (Naim et al., 2021).

7.4 PEROVSKITE PHOTOCELLS

The perovskite photovoltaic technology is the newest among the emerging third generation of solar devices. However, the progress in the field of perovskite solar cells in the subsequent years has led to a certified efficiency exceeding 22%, attracting the interest of potential producers. Table 7.2 presents the photovoltaic parameters of PSC measured by certified laboratories under Standard Test Conditions (STC) in recent years. Results are reported for perovskite single cell, minimodule or submodule. The year 2019 is omitted, since the obtained data are the same as in 2018. The certified efficiency increase of around 3% was registered from 2016 to 2021.

Although the perovskite solar cells achieved relatively high efficiencies, they suffer from instability that hinders commercialization of this promising technology. Thus, a lot of effort has been devoted to upscaling of the PSC devices accompanied by long-term retaining efficiency.

Table 7.2 Photovoltaic parameters of perovskite technology measured by certified laboratories under Standard Test Conditions

Type of the cell	J_{SC} (mA/cm^2)	V_{OC} (V)	FF (%)	Efficiency (%)	Year of publication
Cell	24.67	1.104	72.3	19.7 ± 0.6	2016
Cell	24.67	1.104	72.3	19.7 ± 0.6	2017
Minimodule	19.51	1.029	76.1	16.0 ± 0.4	2017
Cell	24.92	1.125	74.5	20.9 ± 0.7	2018
Minimodule (seven serial cells)	20.66	1.070	78.1	17.25 ± 0.6	2018
Submodule (44 serial cells)	14.36	1.073	75.8	11.7 ± 0.4	2018
Cell	21.64	1.193	83.6	21.6 ± 0.6	2020
Minimodule (seven serial cells)	21.53	1.070	78.4	18.0 ± 0.6	2020
Cell	22.73	1.178	84.4	22.6 ± 0.6	2021
Minimodule (12 cells)	23.09	1.155	75.4	20.1 ± 0.4	2021

Source: Data from Green et al. (2016, 2017, 2018, 2020, 2021b).

The important step toward commercialization was the development of the perovskite cell with organic charge transport layers deposited on non-expensive, flexible foils (Malinkiewicz et al., 2014). On the basis of this invention, the Saule Technologies® company started to operate aiming at the production of thin-film perovskite modules for BIPV and Internet of things (IoT) market. In 2021, the implementation of solar blinds with perovskite cells mounted outdoor on building facade was demonstrated in Poland (Saule, 2022).

The manufacturing of flexible photovoltaic devices requires the utilization of flexible electrodes characterized by high conductivity and transparency. The development and application of such electrodes made of PEDOT:PSS with fluorosurfactant dopant was demonstrated in PCS modules of 25 cm^2 aperture area and 10.9% PCE (Hu et al., 2019).

7.5 QUANTUM DOT-SENSITIZED PHOTOCELLS

At early stage, quantum dot solar cells presented poor efficiency due to low load of QDs and instability. However, the enhancement in efficiency due to the use of multiple exciton generation phenomenon accompanied by low fabrication cost of QDSSCs gradually attracted the attention of not only scientists, but also the companies developing solar technologies. Nowadays, QDSSCs find application as coatings of glass in smart windows that can support the power supply of indoor devices. There are companies that implemented the glass laminated with a layer containing quantum dots, which is intended for use in smart buildings (UbiqD, 2022). The photoactive coating of glass can absorb UV and IR, thus filtering UV and limiting the heating of building interior (MLSystem, 2022).

7.6 INDOOR APPLICATIONS OF EMERGING PHOTOVOLTAICS

Recently, the indoor applications of solar cells have attracted much attention due to the demand for power sources, which can provide electricity for low-consumption wireless electronic devices, especially related to rapidly developing Internet of things (IoT). Additionally, the use of indoor photovoltaics integrated with devices is beneficial for battery lifetime, since it can charge a device even if it is not currently in use.

The experimental studies demonstrated that the third generation solar cells provide efficiencies exceeding 25% when applied indoor, under ambient light, and thus exhibit tremendous potential in this area. Moreover, the indoor performance of emerging photovoltaic cells is better than known for high-efficiency inorganic solar cells based on III–V compounds whose PCE reaches 21% (Biswas & Kim, 2020).

The exceptional performance with artificial indoor lighting was demonstrated by all types of third generation solar cells, including OPV. The optimized deposition of organic photoactive layer enabled the achievement of a power density of 78.2 μW/cm^2 and PCE of 28% under 1000 lux fluorescent light (Lee et al., 2018). Indoor operation of OPV modules with great stability of PCE that was retained for 9 months was also demonstrated. The solution-processed encapsulated roll-to-roll produced modules exhibited a high PCE of 18% under 400 lux LED illumination. The results showed that indoor operation enables using OPV modules as a reasonable electric power source for small electronic devices (Miranda et al., 2021).

Dye-sensitized cells also surpassed inorganic cells under artificial light, after the modification of their structure in which the TiO_2/dye electrode and hole transporting PEDOT were in direct contact (Cao et al., 2018). The beneficial performance was achieved due to reduction in diffusion path of the redox mediator, which resulted in resistance decrease. The PCE of 32% was obtained by using 1000 lux fluorescent light compared to 13.1% under 1000 W/cm^2 and AM1.5G solar light. The PCE of indoor DSSC was then enhanced up to 34% at 1000 lux through the co-sensitization strategy. Two dyes (coded as XY1 and L1 dye) covering the 300–700 nm absorption range and copper (II/I) electrolyte were used in the cells assembled into an array with 16 cm^2 active area. Under fluorescent light, the dye cells demonstrated the potential to power the prototype of autonomous artificial intelligence device (Michaels et al., 2020).

The record of indoor performance efficiency exceeding 36% accompanied by good stability was achieved by the perovskite cell of inverted planar structure under 1000 lux fluorescent light. The utilized perovskite material $CH_3NH_3PbI_{2-x}BrCl_x$ exhibited an extremely low trap-state density and a bandgap of 1.8 eV, which was well matched to the spectrum of the ambient light source (Cheng et al., 2019).

In industrial scale, the demand for supplying electricity to indoor devices was addressed by the Fujikura® company, which produces high-efficiency, reliable DSSC modules of different sizes (from 39.9 mm×35 mm to 70 mm×92 mm) that provide 42–340 μW of power under 200 lux white LED illumination (Fujikura, 2022). Another company, Ricoh®, developed solid-state 30 cm×30 cm DSSC modules that are capable of operating under scattered or indoor light. The submodule with dimensions 5.2 cm×8.4 cm has a maximum power output of 230 μW (Ricoh, 2022).

7.7 TANDEM PHOTOCELLS

The beneficial option that enables overcoming the theoretical efficiency limit for single-junction cell is a tandem (or multi-junction) PV device, which provides the absorption in a wide range of photon energies, owing to the integration of bottom cell based on narrow-bandgap semiconductor with a wide-bandgap top cell. This approach limits thermalization and transmission losses, which typically occur at the single junction of semiconductors due to the differences between incident photon energy and bandgap energy. The theoretical limit of power conversion efficiency of tandem solar cells is around 47%.

Metal halide perovskites are popular materials utilized in the tandem structures, e.g., perovskite-silicon, perovskite-CIGS, perovskite-OPV, perovskite/QDs or wide-bandgap perovskite with narrow-bandgap perovskite (Yeom et al., 2020).

A good example is a multilayer structure of tandem perovskite/Si cell containing amorphous Si, monocrystalline Si and perovskite photoactive absorber with the bandgap of 1.68 eV. On the both sides of the tandem structure, Ag contacts were made. The introduction of modified hole-selective layer in such a cell resulted in the acceleration of the hole transport and reduction in non-radiative charge carriers recombination. A PCE of over 29% was achieved, and 95% of initial efficiency was maintained for 300 h (Al-Ashouri et al., 2020). According to simulation results, the tandem cell consisted

Table 7.3 Photovoltaic parameters of tandem PV cells containing third generation subcells

Top/bottom cell	Structure	J_{SC} (mA/cm^2)	V_{OC} (V)	FF (%)	PCE (%)	References
PSC/CIGSe, 1 cm^2	LiF/Ag/IZO/SnO$_2$/C$_{60}$/Cs$_{0.05}$(MA$_{0.17}$FA$_{0.83}$)$_{0.95}$Pb(I$_{0.83}$Br$_{0.17}$)$_3$/MeO-2PACz/AZO/i-ZnO/CdS/RbF/CIGS/Mo	19.17	1.68	71.9	23.26	Al-Ashouri et al. (2019)
PSC/CIGSe, 0.8 cm^2	LiF/IZO/SnO$_2$/C$_{60}$/Cs$_{0.05}$(MA$_{0.17}$FA$_{0.83}$)Pb$_{1.1}$(I$_{0.83}$Br$_{0.17}$)$_3$/PTAA/NiO$_x$/AZO/i-ZnO/CdS/CIGS/Mo	18.0	1.58	76.0	21.6	Jošt et al. (2019)
PSC/OPV	ITO/PEDOT:PSS/MAPbI$_3$/PCBM/C$_{60}$-SB/Ag/MoO$_3$/PCE-10:PC$_{71}$BM/C$_{60}$-N/Ag	13.1	1.63	75.1	16	Liu et al. (2016)
PSC/OPV	ITO/NiO$_x$/PSS/FAPbBr$_{2.43}$Cl$_{0.57}$/PSS/PCBM/ZnO-NP/LS-ITO/PEDOT:PSS/PTB7-Th:6TIC-4F/ZnO-NP/ITO	6.6	2.23	74.0	10.7	Zuo et al. (2019)
PSC/QD	ITO/c-TiO$_2$/mp-TiO$_2$/MAPbI$_3$/PTAA/MoO$_x$/Au/ZnO/PbS CQD/MoO$_x$/Au	5.0	0.98	20.0	1.01	Karani et al. (2018)
DSSC/CIGS	FTO/TiO$_2$/N719 dye/Pt/i-ZnO/AZO/CdS/CIGS/Mo	14.6	1.17	77	13	Moon et al. (2015)
DSSC/GGC	FTO/TiO$_2$/D131 dye/FTO/glass/p-GaAs/GaAs/Al$_x$Ga$_{(1-x)}$As/n-GaAs	4.99	1.85	82.6	7.63	Ito et al. (2011)
OPV/OPV	ITO/ZnO/PFN-Br/PBDB-T:F-M/M-PEDOT/ZnO/PTB7-Th:O6T-4F:PC$_{71}$BM/Mo$_x$/Ag	14.35	1.64	73.37	17.3	Meng et al. (2018)
OPV/OPV	ITO/PEDOT:PSS/PBDB-T-2F:TfIF-4FIC/PF3N-2TNDI/Ag/PEDOT:PSS/PTB7-Th:PCDTBT:IEICO-4F/PF3N-2TNDI/Ag	13.6	1.6	69	15	Liu et al. (2019)

of 1000 nm thick perovskite with a bandgap of 1.66 eV and the textured Si layer can achieve a PCE of 32.5% (Jošt et al., 2018). The rapid development of tandem perovskite/Si cells at laboratory stage drives the industrial deployment of such cells. Two companies, TNO® and Soliance®, established international cooperation aiming at the transfer of laboratory results in this regard into marketable technologies (TNO, 2022). The main goal is to integrate thin, efficient tandem modules with vehicles, buildings and roads.

The perovskites used with organic materials can constitute a semitransparent tandem PSC/OPV cell, which demonstrates selective absorption of light in the subsequent layers. The exemplary device based on ultra-large-bandgap perovskite (bandgap of 2.36 eV) and bulk heterojunction blend of organic semiconductors with a band edge around 980 nm provided a PCE of 10.7% (Zuo et al., 2019).

In another approach, the combination perovskites and quantum dots was implemented in tandem solar cell. The incorporation of the two materials enabled significant improvement in the solar light harvesting, since perovskites absorb in UV-Vis range and quantum dots are tuned to the absorption of NIR wavelengths. Such a tandem structure was applied in flexible, lightweight modules manufactured in the roll-to-roll process, the long-term durability tests of which are ongoing (QDSolarinc, 2022).

Tandem photovoltaics is a great option, since it provides a lot of room for improvement in terms of the materials the subcells are built of. Third generation technologies can be used in a variety of configurations, which is shown in Table 7.3 that presents structures and performance parameters of different exemplary tandem cells.

7.8 SUMMARY

Significant progress and achievements in the development of third generation photovoltaic cells show their potential in the context of clean energy production. Nevertheless, the overall performance, especially the efficiency and stability, needs to see improvement before the emerging PV technologies will be competitive to well-adopted silicon modules. The current development stage of third generation of photocells indicates that BIPV and niche applications, such as power supply for portable electronic devices or stand-alone systems, will gain the most popularity. Undoubtedly, the demand for powering of electronic devices in various locations and environments, both outdoor and indoor, is expected to increase, inter alia, due to the prospective development of Internet of things and sustainable buildings.

REFERENCES

Al-Ashouri, A., Magomedov, A., Roß, M., Jošt, M., Talaikis, M., Chistiakova, G., Bertram, T., Márquez, J.A., Köhnen, E., Kasparavičius, E., Levcenco, S., Gil-Escrig, L., Hages, C.J., Schlatmann, R., Rech, B., Malinauskas, T., Unold, T., Kaufmann, C.A., Korte, L., Niaura, G., Getautis, V. & Albrecht, S. (2019) Conformal monolayer contacts with lossless interfaces for perovskite single junction and monolithic tandem solar cells. *Energy and Environmental Science* 12(11), 3356–3369. doi: 10.1039/C9EE02268F.

Al-Ashouri, A., Köhnen, E., Li, B., Magomedov, A., Hempel, H., Caprioglio, P., Márquez, J.A., Morales Vilches, A.B., Kasparavicius, E., Smith, J.A., Phung, N., Menzel, D., Grischek, M., Kegelmann, L., Skroblin, D., Gollwitzer, C., Malinauskas, T., Jošt, M., Matic, G., Rech, B.,

Schlatmann, R., Topic, M., Korte, L., Abate, A., Stannowski, B., Neher, D., Stolterfoht, M., Unold, T., Getautis, V. & Albrecht, S. (2020) Monolithic perovskite/silicon tandem solar cell with >29% efficiency by enhanced hole extraction. *Science* 370(6522), 1300–1309. doi: 10.1126/science.abd4016.

Anctil, A. & Fthenakis, V. (2012). Life cycle assessment of organic photovoltaics. In: Fthenakis, V. (ed.), *Third Generation Photovoltaics*. IntechOpen: London. doi: 10.5772/38977; https://www.intechopen.com/chapters/32591.

Barichello, J., Vesce, L., Mariani, P., Leonardi, E., Braglia, R., Di Carlo, A., Canini, A. & Reale, A. (2021) Stable semi-transparent dye-sensitized solar modules and panels for greenhouse application. *Energies* 14, 6393. doi: 10.3390/en14196393.

Biswas, S. & Kim, H. (2020) Solar cells for indoor applications: progress and development. *Polymers* 12, 1338. doi: 10.3390/polym12061338.

Cao, Y., Liu, Y., Zakeeruddin, S.M., Hagfeldt, A. & Grätzel, M. (2018) Direct contact of selective charge extraction layers enables high-efficiency molecular photovoltaics. *Joule* 2(6), 1108–1117. doi: 10.1016/j.joule.2018.03.017.

Cheng, R., Chung, C.-C., Zhang, H., Liu, F., Wang, W.-T., Zhou, Z., Wang, S., Djurišić, A.B. & Feng, S.-P. (2019) Tailoring triple-anion perovskite material for indoor light harvesting with restrained halide segregation and record high efficiency beyond 36%. *Advanced Energy Materials* 9(38), 1901980. doi: 10.1002/aenm.201901980.

Fujikura (2022) Available from: https://www.fujikura.co.jp/eng/newsrelease/products/2062445_11777.html [Accessed March 22, 2022].

G24 Manufacturing Process (2022) Available from: https://gcell.com/about-g24-power/manufacturing-process [Accessed March 22, 2022].

Green, M.A., Emery, K., Hishikawa, Y., Warta, W. & Dunlop, E.D. (2016) Solar cell efficiency tables (version 48). *Progress in Photovoltaics: Research and Applications* 24(7), 905–913. doi: 10.1002/pip.2788.

Green, M.A., Hishikawa, Y., Warta, W., Dunlop, E.D., Levi, D.H., Hohl-Ebinger, J. & Ho-Baillie, A.W.H. (2017) Solar cell efficiency tables (version 50). *Progress in Photovoltaics: Research and Applications* 25(7), 668–676. doi: 10.1002/pip.2909.

Green, M.A., Hishikawa, Y., Dunlop, E.D., Levi, D.H., Hohl-Ebinger, J., Anita, W.Y. & Ho-Baillie, A.W.Y. (2018) Solar cell efficiency tables (version 52). *Progress in Photovoltaics: Research and Applications* 26(7), 427–436. doi: 10.1002/pip.3040.

Green, M.A., Dunlop, E.D., Hohl-Ebinger, J., Yoshita, M., Kopidakis, N. & Hao, X. (2020) Solar cell efficiency tables (version 56). *Progress in Photovoltaics: Research and Applications* 28(7), 629–638. doi: 10.1002/pip.3303.

Green, M., Dunlop, E., Hohl-Ebinger, J., Yoshita, M., Kopidakis, N. & Hao, X. (2021a) Solar cell efficiency tables (version 59). *Progress in Photovoltaics* 30(1), 3–12. doi: 10.1002/pip.3506.

Green, M.A., Dunlop, E.D., Hohl-Ebinger, J., Yoshita, M., Kopidakis, N. & Hao, X. (2021b) Solar cell efficiency tables (version 58). *Progress in Photovoltaics: Research and Applications* 29(7), 657–667. doi: 10.1002/pip.3444.

Hu, X., Meng, X., Zhang, L., Zhang, Y., Cai, Z., Huang, Z., Su, M., Wang, Y., Li, M., Li, F., Yao, X., Wang, F., Ma, W., Chen, Y. & Song, Y. (2019) A mechanically robust conducting polymer network electrode for efficient flexible perovskite solar cells. *Joule* 3(9), 2205–2218. doi: 10.1016/j.joule.2019.06.011.

Ito, S., Dharmadasa, I.M., Tolan, G.J., Roberts, J.S., Hill, G., Miura, H., J.-H. Yum, J.-H., Pechy, P., Liska, P., Comte, P. & Grätzel, M. (2011) High-voltage (1.8V) tandem solar cell system using a GaAs/Al$_x$Ga$_{(1-x)}$As graded solar cell and dye-sensitised solar cells with organic dyes having different absorption spectra. *Solar Energy* 85(6), 1220–1225. doi: 10.1016/j.solener.2011.02.024.

Jošt, M., Bertram, T., Koushik, D., Marquez, J.A., Verheijen, M.A., Heinemann, M.D., Köhnen, E., Al-Ashouri, A., Braunger, S., Lang, F., Rech, B., Unold, T., Creatore, M.,

Lauermann, I., Kaufmann, C.A., Schlatmann, R. & Albrecht, S. (2019) 21.6%-efficient monolithic perovskite/Cu(In, Ga)Se$_2$ tandem solar cells with thin conformal hole transport layers for integration on rough bottom cell surfaces. *ACS Energy Letters* 4, 583–590. doi: 10.1021/acsenergylett.9b00135.

Jošt, M., Köhnen, E., Morales-Vilches, A.B., Lipovšek, B., Jäger, K., Macco, B., Al-Ashouri, A., Krč, J., Korte, L., Rech, B., Schlatmann, R., Topič, M., Stannowski, B. & Albrecht, S. (2018) Textured interfaces in monolithic perovskite/silicon tandem solar cells: advanced light management for improved efficiency and energy yield. *Energy Environmental Science* 11(12), 3511–3523. doi: 10.1039/C8EE02469C.

Karani, A., Yang, L., Bai, S., Futscher, M.H., Snaith, H.J., Ehrler, B., Greenham, N.C. & Di, D. (2018) Perovskite/colloidal quantum dot tandem solar cells: theoretical modeling and monolithic structure. *ACS Energy Letters* 3, 869–874. doi: 10.1021/acsenergylett.8b00207.

Krebs, F.C., Tromholt, T. & Jørgensenn, M. (2010) Upscaling of polymer solar cell fabrication using full roll-to-roll processing. *Nanoscale* 2(6), 873–886. doi: 10.1039/B9NR00430K.

Lee, H.K.H., Wu, J., Barbé, H., Jain, S.M., Wood, S., Speller, E.M., Li, Z., Castro, F.A., James R. Durrant, J.R. & Tsoi, W.C. (2018) Organic photovoltaic cells: promising indoor light harvesters for self-sustainable electronics. *Journal of Materials Chemistry A* 6, 5618–5626. doi: 10.1039/C7TA10875C.

Lee, C.-Y., Tsao, C.-S., Lin, H.-K., Cha, H.-C., Chung, T.-Y., Sung, Y.-M. & Huang, Y.-C. (2021) Encapsulation improvement and stability of ambient roll-to-roll slot-die-coated organic photovoltaic modules. *Solar Energy* 213, 136–144. doi: 10.1016/j.solener.2020.11.021.

Liu, Y., Renna, L.A., Bag, M., Page, Z.A., Kim, P., Choi, J., Emrick, T., Venkataraman, D. & Russell, T.P. (2016). High efficiency tandem thin-perovskite/polymer solar cells with a graded recombination layer. *ACS Applied Materials Interfaces* 8, 7070–7076. doi: 10.1021/acsami.5b12740.

Liu, G., Jia, J., Zhang, K., Jia, X., Yin, Q., Zhong, W., Li, L., Huang, F. & Cao, Y. (2019) 15% efficiency tandem organic solar cell based on a novel highly efficient wide-bandgap nonfullerene acceptor with low energy loss. *Advanced Energy Materials* 9(11), 1803657. doi: 10.1002/aenm.201803657.

Lucera, L., Machui, F., Schmidt, H.D., Ahmad, T., Kubis, P., Strohm, S., Hepp, J., Vetter, A., Egelhaaf, H.-J. & Brabec, C.J. (2017) Printed semi-transparent large area organic photovoltaic modules with power conversion efficiencies of close to 5%. *Organic Electronics* 45, 209–214. doi: 10.1016/j.orgel.2017.03.013.

Malinkiewicz, O., Yella, Y., Lee, Y.H., Espallargas, G.M., Graetzel, M., Nazeeruddin, M.K. & Bolink, H.J. (2014) Perovskite solar cells employing organic charge-transport layers. *Nature Photonics* 8, 128–132. https://www.nature.com/articles/nphoton.2013.341.

Meng, L., Zhang, Y., Wan, X., Li, C., Zhang, X., Wang, Y., Ke, X., Xiao, Z., Ding, L., Xia, R., Yip, H.-L., Cao, Y. & Chen, Y. (2018) Organic and solution-processed tandem solar cells with 17.3% efficiency. *Science* 361(6407), 1094–1098. doi: 10.1126/science.aat2612.

Michaels, H., Rinderle, M., Freitag, R., Benesperi, I., Edvinsson, T., Socher, R., Gagliardi, A. & Freitag, M. (2020) Dye-sensitized solar cells under ambient light powering machine learning: towards autonomous smart sensors for the internet of things. *Chemical Science* 11, 2895–2906. doi: 10.1039/C9SC06145B.

Miranda, B.H.S., de Q. Corrêa, L., Soares, G.A., Martins, J.L., Lopes, P.L., Vilela, M.L., Rodrigues, J.F., Cunha, T.G., de Q. Vilaça, R., Castro-Hermosa, S., Wouk, L. & Bagnis, D. (2021) Efficient fully roll-to-roll coated encapsulated organic solar module for indoor applications. *Solar Energy* 220, 343–353. doi: 10.1016/j.solener.2021.03.025.

MLSystem (2022) Available from: https://mlsystem.pl [Accessed March 30, 2022].

Moon, S.H., Park, S.J., Kim, S.H., Lee, M.W., Han, J., Kim, J.Y., Kim, H., Hwang, Y.J., Lee, D.-K. & Min, B.K. (2015) Monolithic DSSC/CIGS tandem solar cell fabricated by a solution process. *Scientific Reports* 5, 8970. doi: 10.1038/srep08970.

Mozaffari, S., Nateghi, M.R. & Zarandi, M.B. (2017) An overview of the challenges in the commercialization of dye sensitized solar cells. *Renewable and Sustainable Energy Reviews* 71, 675–686. doi: 10.1016/j.rser.2016.12.096.

Naim, W., Novelli, V., Nikolinakos, I., Barbero, N., Dzeba, I., Grifoni, F., Ren, Y., Alnasser, T., Velardo, A., Borrelli, R., Haacke, S., Zakeeruddin, S.M., Graetzel, M., Barolo, C. & Sauvage, F. (2021) Transparent and colorless dye-sensitized solar cells exceeding 75% average visible transmittance. *Journal of American Chemical Society Au* 1(4), 409–426. doi: 10.1021/jacsau.1c00045.

QDSolarinc (2022) Available from: https://qdsolarinc.com/ [Accessed February 15, 2022].

Ricoh (2022) Available from: https://www.ricoh.com/technology/tech/066_dssc [Accessed February 2, 2022].

Saule (2022) Available from: https://sauletech.com/ [Accessed March 30, 2022].

Solaronix (2022) Available from: https://www.solaronix.com/solarcells/applications/ [Accessed February 15, 2022].

Suzuki, K., Izawa, S., Chen, Y., Nakano, K. & Tajima, K. (2017) Drawing organic photovoltaics using paint marker pens. *AIP Advances* 7(11), 115002. doi: 10.1063/1.5006352.

TNO (2022) Available from: https://www.tno.nl/en/focus-areas/energy-transition/roadmaps/renewable-electricity/solar-energy/solar-panel-efficiency/solliance/ [Accessed January 22, 2022].

UbiqD (2022) Available from: https://ubiqd.com/solar/ [Accessed February 4, 2022].

Weerasinghe, H.C., Vak, D., Robotham, B., Fell, C.J., Jones, D. & Scully, A.D. (2016) New barrier encapsulation and lifetime assessment of printed organic photovoltaic modules. *Solar Energy Materials and Solar Cells* 155, 108–116. doi: 10.1016/j.solmat.2016.04.051.

Wikipedia (2022) Available from: https://en.wikipedia..org/wiki/Konarka_Technologies [Accessed February 4, 2022].

Yeom, K.M., Kim, S.U., Woo, M.Y., Noh, J.H. & Im, S.H. (2020) Recent progress in metal halide perovskite-based tandem solar cells. *Advanced Materials* 32(51), 2002228. doi: 10.1002/adma.202002228.

Zuo, L., Shi, X., Fu, W. & Jen, A.K.-Y. (2019) Highly efficient semitransparent solar cells with selective absorption and tandem architecture. *Advanced Materials* 31(36), 1901683. doi: 10.1002/adma.201901683.

Index

absorption of light 8, 27, 48, 77, 105
applications of third generation photocells 139
artificial lighting 141–142

bandgap 10–11, 16, 77–78, 80, 104, 107
building integrated photovoltaics 128

commercialization 138, 141
counter electrode 59–60, 115
    conductive polymers 60
    nanocarbon 60
    platinum black 59
    recycling 122
cumulative energy demand 120, 129, 134

downconversion 11–12
dye 48, 56–59
    absorption 56, 59
    excited state 50
    metal-free 58
    natural 59
    porphyrine 57
    ruthenium complex 57
dye-sensitized cell 47–53
    all-solid-state 62
    charge transfer 50
    inverted 52
    life cycle assessment 127–130
    manufacturing 139
    mesoporous TiO_2 48–49, 53–56
    performance 56–57, 61
    photoanode 48–49
    recombination of carriers 51
    sensitization 48
    structure 48

electrolyte 49, 60
    copper redox couple 61
    gel electrolyte 62, 114
    iodide/triiodide 61
    polysulfide liquid electrolyte 114
electron transport layer 83, 113
    inorganic 84, 113
    nanocarbon in ETL 85, 114
    organic 84
energy payback time 120, 124, 131, 134
energy return factor 121, 124
environmental performance 119–120
experimental installation 128, 138, 139, 141

flexible module 138, 141

global warming potential 121
greenhouse gas emissions 121, 125, 127, 129, 131

hole transport layer 130, 85
    inorganic 62
    nanocarbon in HTL 86
    organic 62, 85–86
HOMO 27, 50

indoor performance of photocells 141–142
internet of things 141

life cycle 119, 123
life cycle assessment 120
lifetime 121
light trapping 14, 37–40
LUMO 27, 50

manufacturing 138
multijunction solar cell 21, 142, 144

net energy ratio 121, 134

organic photocell 26–40
    acceptor 30
    blend morphology 34
    bulk heterojunction 29–30
    commercialization 138
    degradation 40
    donor 30
    energy levels 29

organic photocell (*cont.*)
  environmental impact 123–127
  ETL 35
  fullerene acceptors 30–31
  fullerene derivatives 30–31
  losses 37
  non-fullerene acceptors 32–33
  organic semiconductors 26, 29
  performance 30–34, 137
  stability 40–41
  structure 28–29, 34
  ternary blend 36

perovskite 69–76
  bandgap 77–78, 80
  energy levels 72
  environmental performance 130–133
  hybrid perovskites 76–79, 91, 92, 131
  inorganic perovskites 70, 73, 79–82, 90, 92
  lead-free perovskites 94
  organometal halide perovskite 69
  phase transitions 73
  structure 72, 74
perovskite cell 76–83
  all-inorganic 85, 86
  commercialization 141
  current-voltage characteristic 88
  degradation 91
  history 69–71
  hysteresis 88–90
  life cycle assessment 130–133
  lifetime 90, 92
  performance 79, 85, 86, 140
  stability 90–93
  structure 82–83
photovoltaic parameters 13, 137, 140, 143
  of perovskite technology 140
  of silicon solar cell 137
  of tandem photocells 142
  of third generation solar cells 137
photoelectrochemical cell 47–48

quantum dots 104–113
  absorption 105, 107–108
  bandgap 104, 107
  composition 104
  core-shell structure 106
  doping 105
  non-toxic 133
  sensitization 110–113
  ternary alloy dots 106
quantum dot-sensitized solar cell 102–110
  charge transfer 103
  co-sensitization 108
  current-voltage characteristic 103
  life cycle 133–134

multiple exciton generation 109
  performance 107, 113, 115
  quantum efficiency 109
  structure 103
quantum efficiency 6, 51–52, 58, 70, 109

recombination of carriers 9, 51
recycling 121–122

semitransparency 36, 138, 139, 144
solar cell operation 1–11
  current-voltage characteristic 4, 52, 88, 91
  efficiency 5, 13, 85, 86, 137, 140
  electrical parameters 3–5
  fill factor 5, 51, 140
  fundamentals 2–3
  light harvesting efficiency 52
  maximum power 5
  open-circuit voltage 3, 5, 29, 51
  operation 2–3, 7
  parasitic resistance 5–6
  quantum efficiency 52
  recent achievements 11
  short-circuit current 4, 51
  silicon 6–7, 11
  two-diode model 5

spectral response 6
stability 40–41

thin-film solar cell 12–21
  amorphous silicon 18
  CdTe 15
  CIGS 16
  GaAs 21
  structure 13
  thin-film silicon 19–20
titanium dioxide 69, 111, 113
transparent conductive oxide 15, 17, 28, 48, 133
  FTO 48, 122, 129
  ITO 133
  recycling 122

upconversion 11–12, 40
upscaling 138, 140